David Wark

Prevention and Cure of Chronic Consumption

David Wark

Prevention and Cure of Chronic Consumption

ISBN/EAN: 9783337729455

Printed in Europe, USA, Canada, Australia, Japan

Cover: Foto ©berggeist007 / pixelio.de

More available books at **www.hansebooks.com**

PREVENTION AND CURE

OF

CHRONIC CONSUMPTION.

BY

DAVID WARK, M.D.,

PROFESSOR OF OBSTETRICS AND DISEASES OF WOMEN AND CHILDREN IN THE
UNITED STATES MEDICAL COLLEGE, NEW YORK.

NEW YORK:
THE AUTHORS' PUBLISHING COMPANY.
1880.

PREFACE.

My object in publishing this monograph is to attract attention to what I believe to be a valuable advance in the treatment of chronic pulmonary consumption, so that those in whom the disease is being developed may receive the benefits it is so well adapted to confer, when prescribed by and carried out under the direction of physicians who are also expert in the use of the treatment here advocated.

Nothing can savor less of quackery or empiricism than this practice, because all its principles are the direct deductions of modern physiological science, and the treatment is the practical use for the cure of disease of those demonstrated truths. Therefore, it possesses claims upon and is thoroughly worthy of the confidence of intelligent and educated seekers of health.

It is, therefore, placed before the public with the conviction that the value of physical training in the prevention and cure of chronic pulmonary consumption, added to the current treatment, only requires to be known, in order to be highly appreciated and extensively employed as a remedy for this and other chronic disorders that are characterized by defective nutrition.

DAVID WARE, M. D.

No. 4 WEST 43d STREET, NEW YORK.

CONTENTS.

	PAGE.
DECAY AND RENEWAL OF THE BODY.	7
HEALTH, DISEASE AND CURE	10
THE EARLY SYMPTOMS OF CONSUMPTION.	12
CATARRHAL CONSUMPTION.	16
FIBROUS CONSUMPTION	17
TUBERCULAR CONSUMPTION—THE ORIGIN OF TUBERCLES	18
THE RATE AT WHICH THE BREATHING POWER DECLINES.	22
HUTCHINSON'S TABLE.	23
THE EFFECT OF INADEQUATE BREATHING ON THE QUALITY OF THE BLOOD.	25
THE CONNECTION OF RESPIRATION WITH THE GENERATION OF POWER	28
IS CONSUMPTION CURABLE.	32
THE CURABLE AND INCURABLE VARIETIES OF CONSUMPTION.	35
ILLUSTRATIVE CASES (six)	36
THE CHIEF DIFFICULTY IN EFFECTING CURES	51
SHOWING WHY AND WHERE THE PREVALENT TREATMENT IS DEFECTIVE	53
THE MECHANISM OF BREATHING.	61
THE EFFECT OF EXERCISE ON HEALTHY PERSONS.	63
THE EFFECTS OF EXERCISE ON INVALIDS.	64
THAT WHICH CONSUMPTIVES MOST URGENTLY NEED	71

PAGE.

What Special Training can do 74

Effects of Special Training on the Breathing
Organs............ 77

The Increased Air-space Usually obtained in the
Lungs by Special Training..................... 78

Showing why Training to be Curative in this Disease must be Special............. 80

Practice of Special Training..................... 82

The Effects of Special Training on the Circulation
of the Blood.................................. 87

the Influence of Special Training on the Production of Vital Heat............................. 89

Review of the Objections sometimes urged against
Special Curative Training..................... 91

    1st Obj.—*As it is applied externally it cannot reach
any internal disease....* 92

    2d Obj.—*Such treatment must be dangerous to per-
sons who have a tendency to hemorrhage.............* 93

    3d Obj.—*Physical training must be harsh and ex-
haustive, therefore, it is unsuited to the treatment of
delicate persons ....* 94

The Advantages of Special Training.—When properly used it never injures..................... 96

The Curative Effects are quickly apparent........ 97

The Action of Medicines with Special Training 98

Permanence of the Results....................... 101

Special Training as a Palliative for Incurable
Cases.... 102

Special Training as a Preventive 103

PREVENTION AND CURE

OF

CHRONIC CONSUMPTION.

In order to make the views I hold concerning the nature and treatment of pulmonary consumption quite intelligible, it will be necessary to previously consider the physiological changes ever going on in the body, and the meaning of the terms Health, Disease and Cure.

DECAY AND RENEWAL OF THE BODY.

Regular and sufficient supplies of water, food and air are essential to the continuance of life. A man requires for this purpose, in the course of a year, more than three thousand pounds of materials, an amount equal to about twenty times his own weight. He consumes eight hundred pounds of solid food, absorbs from the atmosphere an equal

(7)

weight of oxygen, and drinks about fifteen hundred pounds of water. An abundant supply of air is needed every moment; food and drink must be taken at frequent intervals. Why is the demand so imperious? The modern physiologist declares it is because the essential condition of life is death. Decay is more truly a part of life than it is of death, because it goes on during all our physical existence, but after dissolution it ceases, when the work of decomposing the organic particles of which the body is made up into inorganic elements has been completed. The living body is like the flame of a lamp, continually fed, but as continually wasting away. It is a noble mansion, built of wonderfully wrought but perishable materials.

The rate at which the living body undergoes destructive change is much more rapid than is popularly supposed. It has been shown that the bodies of mammals contain about 35.45 grammes of nitrogen per kilogramme, and therefore the body of a man who weighs one hundred and thirty pounds contains upwards of 4.6 pounds of nitrogen. Baron Liebig, moreover, states that the liquid and solid excreta of a man by kidneys and bowels for a year, contain 16.41 pounds of nitrogen, or for three months and a half 4.7 pounds of nitrogen. From these data we conclude that all the soft parts of a healthy human body are removed by vital change, and completely renewed from the nutritive materials consumed as food, air and water, in about three months and a half. On the other hand, experiments conducted at the morgue in Paris show that the features retain for three and a half months the aspect they wore during life, so that the individual may be recognized and the age told ; and another month is required before the features become unrecognizable, and the bulky muscles of the trunk and thighs yield to decay. Dead flesh and living flesh are thus equally perishable.

Modern science has exposed the fallacies of the old physiologists who believed that what they called the vital principle endowed the body with the power to resist change. On the contrary, the truth is, the living human organism submits to unceasing waste, and if proper supplies are withheld, it soon perishes. Thus it is absolutely true that our bodies are ever dying. The difference between this ever-occurring death and final dissolution consists in the fact that, in the first instance, no sooner does one atom of a healthy body die, than a living atom is supplied from the blood to take its place and perform its functions. In the second instance the whole fabric returns to dust as it was.

HEALTH, DISEASE AND CURE.

Animal life is carried on by virtue of certain internal functions—Respiration, Digestion, Absorption, Circulation, Assimilation, Secretion and Excretion. Now when the food, water and oxygen that the body requires for its support are subjected to these processes, they undergo two separate and distinct series of changes, progressive and retrogressive. In the first series the nutritive materials are digested, absorbed into the blood, and then become incorporated with the body, so that the food we consumed a few hours before, now forms part of the organs by which we see, think, hear, feel and move. After these materials have remained in the body a certain length of time as living flesh, they begin to pass down through retrogressive changes, by which they are finally fitted to be expelled from the body as inorganic, waste matter. Now, when these two series of vital changes occur perfectly in regular order, the result is perfect nutrition or absolute health ; when they are carried on imperfectly or irregularly, the result is imperfect nutrition or disease in some form. The complete cessation of vital decay and renewal is the death of the whole structure, the partial cessation of these, during life, is disease. After death the body decays very much as it does during life, but renewal is at an end. In disease both these processes are slow and imperfect. The removal of pulmonary consumption, or indeed chronic disease of any kind, therefore demands that the means employed be capable of bringing about, physiologically, the rapid decay and complete renewal of the body.

The effective power of an army is maintained by retiring soldiers before age has deprived them of manly vigor, and supplying their places with younger men. The health and strength of the human body is to be promoted in very much the same way. In order to keep our bodies in vigorous condition, a large proportion of the living atoms of which they are composed must possess the vital properties of recently organized tissue. There must not be doing duty in the body an undue percentage of particles on the downward career of what, for the sake of illustration, may be called atomic old age.

THE EARLY SYMPTOMS OF CONSUMPTION.

The early symptoms of this formidable disease are too often overlooked, or at least attributed to other than their real causes. In fact, the sufferer usually and unwisely ignores them as long as possible.

It is astonishing how often consumptives succeed in blinding themselves to the true nature of their disease. They may be frequently observed well advanced in the second stage, still imagining that all their difficulties are caused by some trifling affection of the throat.

Every case has its peculiarities, both in the beginning and during its progress. The symptoms, however, generally arise as follows :

At first slight, shifting pains are usually felt in the lungs, often amounting only to a feeling of uneasiness. A sense of tightness across the chest is experienced. The breath becomes shortened. This is particularly noticed when any extra muscular exertion is attempted. There is often a sense of chilliness during the fore part of the day, with a feeling of feverishness toward evening. The circumference of the chest slightly diminishes ; its walls lose their elasticity. The collar bones become more prominent. Flesh is slowly but steadily lost in most cases. Digestion is less vigorous than formerly. Nausea is sometimes present. The appetite falls off. A very slight tickling cough exists on rising in the morning. The pulse becomes habitually more frequent. Bleeding from the lungs is common ; in many cases, however, this never occurs

from first to last.   In women an early symptom is
often the gradual cessation of the menses.

Consumptives usually date the beginning of their
disease from the time a cold, real or imaginary,
was caught, an opinion that is sometimes right but
frequently wrong.   An ordinary cold rarely gives
rise to pulmonary consumption in an individual
who has a thoroughly sound pair of lungs.   When
a man whose breathing organs are quite healthy,
takes a cold that "settles on his lungs," he suffers
more or less from such symptoms as lassitude, muscular pains, backache, headache, sore throat, hoarse-
ness, feverishness, thirst, loss of appetite, water runs
from his eyes and his nose, he coughs hard and ex-
pectorates profusely.   These all pass away in a few
days, usually with little or no active treatment,
and if the cough is somewhat obstinate it is readily
cured by a good cough mixture.   But the approach
of this disease is most insidious ; it is not heralded
by any of the above well marked symptoms.   A
consumptive cough, as before stated, commonly begins as a slight dry hack on getting out of bed, or

perhaps it is excited at first only by leaving a warm room and going out into the cold air. Afterwards a little watery or gluey matter is raised ; this gradually becomes thick, heavy, yellow and copious.

If a young adult has a cough of this character, with wandering pains through the chest, and loses flesh, even slightly, he is, in all probability, consumptive. If besides these, he has raised bright red liquid blood, even in small quantities, he is almost certainly so. As a cough of this character is caused by deposits of adventitious matter or inflammation, or by both of these together in some part of the lungs, it is never permanently improved by cough medicines, but commonly goes from bad to worse in spite of the most skillfully prepared mixtures. The symptoms that characterize the later stages are too well known to require mention. A few of the earlier indications only are stated ; to these the reader's attention is directed.

In order to convey to the non-medical reader as accurate a conception of the disease under consideration as possible with the space at our disposal, I shall briefly describe the destructive changes occurring in the lungs during its progress, and afterward notice the mechanical and vital defects that seem to be the main causes of its beginnings, and which, if not effectually remedied, usually carry on the disorder to a fatal termination.

The anatomical changes taking place in the lungs of consumptives have induced pathologists to divide this disease into three varieties, the Catarrhal, the Fibrous, and the Tubercular.

CATARRHAL CONSUMPTION

Is preceded in the large majority of cases by inflammation of the bronchial mucus membrane. This catarrh gradually extends from the larger to the smaller tubes, until it finally reaches those of the smallest size. The passage of air into the vesicles receiving their supply through the inflamed tubes is obstructed, and as a direct result they collapse, that portion of the lung becoming more

or less solidified. The inflammation, when it goes
on a certain time, gives rise to cheesy deposits in
the lung. Under favorable conditions these deposits
may become enclosed in a capsule and the patient
recover. Recoveries from catarrhal consumption
frequently occur under suitable treatment at this
early stage.

FIBROUS CONSUMPTION.

The lungs, like other soft parts, are held together
by areolar or connective tissue. The fibrous form
of consumption is at first an inflammation of this
tissue, the result of which is to cause contraction
of the bronchial tubes and air vesicles. The ingress
of inspired air is thus at first obstructed, and, as the
disease progresses, completely prevented; this process spreads from one part of the lung to another,
until the breathing capacity is so diminished that
life is no longer possible.

TUBERCULAR CONSUMPTION.—THE ORIGIN OF

TUBERCLES.

When food is received into the stomach, although inside the body, it is still truly external to the animal system—the scene of life. Before it can get there and become conducive to bodily nutrition it must pass into and become part of the blood. Nutritive matters that are soluble in water readily find their way from the stomach or intestines through the coats of blood vessels. But food that is soluble only in the digestive juices, like bread or beef, finds its way into the circulation by a more circuitous route. After digestion it is taken up by the absorbents and conducted into the blood. While the digested food is still in the alimentary organs, although it has undergone very important changes during digestion, it is still only dead matter. Vitalization does not begin until after it has been received into the lymphatic vessels ; while passing through these and the mesenteric glands it seems to become progressively endowed with life ; leaving

the absorbent system it enters and mixes with the
current of venous blood ever pouring to the lungs,
where the vitalizing process is completed by expo-
sure to the respired air. Thus the food that was
eaten a few hours before now becomes rich, red
arterial blood, if everything has gone on properly.

All the vital changes that food undergoes in becoming living blood, whether these changes occur
in the lymphatics, the mesenteric glands, the liver
or the lungs—they all require the presence of an
abundant supply of oxygen. A definite quantity
of this vital gas is needed to complete the vitalization of a given quantity of food. A man requires
about two pounds of solid food per day, and very
nearly the same weight of oxygen. Therefore, we
shall not be far from the truth when we say that an
atom of food requires to be acted on in the body
by an atom of oxygen in order that its complete
vitalization may be effected. If the supply of
oxygen in the system is deficient, some portion of
the food must either partially or not at all undergo
the needful vital changes. But the course of this

imperfectly elaborated material cannot be stayed. In order to preserve its own proper constitution the blood cannot retain the nutritive matter which it is constantly receiving from the digestive apparatus ; this must pass from the blood into the solid parts to supply, as best it may, the waste of muscle, brain, nerves, etc. It is therefore evident that the materials furnished from the blood for building up the bodies of persons who breathe too little, are badly fitted for their duties. The tissues are worn out, and they are renewed imperfectly by matter possessing a low degree of vitality, which in consequence of imperfect elaboration has failed to reach the high organization of truly living matter. Some portions of it are so inadequately endowed with life that they cannot be used at all in the renewal of the body ; such matter is therefore deposited from the blood in various parts of the system as tubercles. When these occur at the base of the brain they give rise to that fatal disease known as hydrocephalus. When they fall on certain glands forming a part of the alimentary apparatus, they produce marasmus, another grave disease, and when they accumulate in the lungs they give rise to pulmonary consumption.

These views concerning the origin of tubercles are strongly corroborated by the following fact : In making post-mortem examinations of persons who died of consumption, tubercles of several different kinds are usually found in the same subject ; some of these having been deposited during the initial stages of the disease, before the breathing power was much impaired, bear evident traces of organ-ization, having attained a low degree of vitality. This variety of tubercle has a tendency to contract and remain in the lungs without doing much injury. But as the disease progressed and the breathing capacity diminished, tubercular matter occurs, evincing less and less organization, and manifesting a tendency to break down, until at last we find masses of crude cheesy tubercle that cause circum-scribed inflammation, become softened and break down almost as soon as deposited. These facts, taken in connection with the immunity from consumption enjoyed by those whose respiratory organs are well developed and properly used, as well as the beneficial effects that are promptly secured to consumptives by any increase of the breathing capacity, I believe, fully justify the statement that *tubercles are the result of defective nutrition directly traceable to inadequate respiratory capacity,* either congenital or acquired ; or, to speak perhaps more plainly, *tubercles are composed of particles of food which have failed to acquire life while undergoing the vital processes, because the person in whom they occur habitually breathed too little fresh air.*

THE RATE AT WHICH THE BREATHING POWER DECLINES.

The following table, exhibiting the gradual decline of the breathing capacity in consumption, is the result of investigations made by Dr. Hutchinson, an eminent English physician. The quantity of air that can be expired after the most complete inspiration he terms the vital volume or vital capacity indicated in cubic inches.

HUTCHINSON'S TABLE.

HEIGHT.			In Health.	In First Stage.	In Second Stage.	In Third Stage.
5 feet	1 inch.		174	117	99	82
5 "	2 "		182	122	103	86
5 "	3 "		190	127	108	89
5 "	4 "		198	133	113	93
5 "	5 "		206	138	117	97
5 "	6 "		214	143	122	100
5 "	7 "		222	149	129	104
5 "	8 "		230	154	131	108
5 "	9 "		238	157	136	112
5 "	10 "		246	165	140	116
5 "	11 "		254	170	145	119
6 "	0 "		262	176	149	123

This table shows that even in the very earliest detectable stage of consumption the breathing power declines nearly one-third, and in the second and third stages the deficiency is still more marked. It is quite common to observe consumptives the circumference of whose chests has diminished from one to three inches since the beginning of their disease. This fact is so evident that a resort to measurement to prove it is scarcely necessary. Observe the chest of a consumptive; it is almost invariably narrow and shrunken; its walls are rigid; they rise and fall very little, even during forced breathing. With a pair of lungs thus cooped up in an unyielding box, it is impossible to get into the blood the quantity of oxygen needed for the vital purposes. Every one knows that the breath of consumptives, in unfavorable cases, becomes shorter and shorter, until at last it ceases altogether. This is his chief difficulty; almost all his other distressing symptoms flow from this. As his respiratory power diminishes, they are aggravated; if by any device it is increased, they promptly improve.

· This lack of breathing capacity,. amounting to

one-half of the whole vital volume, has a profoundly malign influence on the system ; it depraves the quality of the blood, prevents its normal and equable circulation, limits the evolution of power and strongly predisposes to the development of chronic diseases of various kinds. The foregoing table indicates the measure of air inhaled during forced breathing ; the amount taken into the lungs in quiet respiration is always much less. But when the capacity of the lungs is much reduced in forced respiration, it has been found that the habitual volume declines in about the same proportion.

THE EFFECT OF INADEQUATE BREATHING ON THE QUALITY OF THE BLOOD.

All the nutritive functions are carried on in the capillaries, not in the large blood vessels. The arteries never communicate directly with the veins, but between them lie minute vessels or interstices. In traversing these the arterial blood becomes changed into venous. These interstices, whether tubular or inter-cellular, constitute the capillary system of the body. The diameter of the vessels and spaces varies from the 1500th to the 10,000th of an inch. The circulation of the blood plasma, chyle, lymph and muscular fluids is altogether independent of the heart's action. This organ has no control over these currents, for it has no connection with the absorbent or capillary system, nor with the interstices between the cells. The office of the heart is to regulate the circulation in the whole body, and to drive the blood into the lungs and arteries ; it does this with a force equal to about one pound to the square inch.

If the web of a frog's foot is exposed in the field of a microscope, the blood will be seen moving leisurely in the capillaries from artery to vein. If the web is touched with a drop of strong brine, the current rushes to the spot, and a centre of inflammation is established. On the contrary, if an acqueous solution of opium is placed on it, the current becomes slower and retires from the part. Heat also increases the rapidity and force of the capillary circulation, and cold has an opposite effect. The blood gives up to the body its nutritive materials, and receives from it the effete products of worn-out atoms, while it is traversing the capillaries. The capillary circulation is, therefore, specially necessary to the nutrition of the body and the purification of the blood.

All the vegetable world, and a majority of the animal creation, are destitute of hearts, yet the capillary circulation of juice and blood in them all is perhaps as perfect as it is in mammals, who are provided with a complicated cardiac apparatus.

The circulation through the capillaries is maintained by a chemical affinity possessed by the blood for the sides of the vessels containing it. The advance of the arterial blood is a phenomenon analogous to that of the ascent of water in a glass tube of small diameter; it is due to the attraction of the fluid and solid for each other. This attraction explains the fact that the arteries are always found empty after death, whilst the veins are filled. At the moment of dissolution, the blood in the arteries is attracted by the sides of the capillaries, hence the vessels that bring it are emptied after the heart has ceased to beat.

But this affinity is possessed in a proper degree only by blood rich in oxygen. When this vital gas is lacking in the blood, it cannot circulate vigorously ; it partially stagnates in the vital organs, and fails to properly reach every part of the body, particularly the skin and the extremities, no matter how hard the heart may beat. Therefore, the blood must circulate irregularly if the breathing capacity is inadequate, because the requisite quantity of oxygen cannot get through the obstructed lungs into the blood to impart to the vital fluid those qualities on which a necessary part of its circulating power depends.

These facts explain why consumptives uniformly suffer from irregular circulation of the vital fluids.

THE CONNECTION OF RESPIRATION WITH THE GENERATION OF POWER.

One of the most important purposes of our bodies

is to generate power—nervous power, muscular power, mental power. There are few things so much coveted as the capacity to evolve power. If this precious commodity could be bottled up and sold like proprietary medicines, its sale would far outstrip that of the most popular remedy. The way in which power is evolved in the human body bears a close resemblance in some points to the way it is generated in a locomotive. Steam, the active agent in the engine, is produced by the burning of fuel. If the fuel in the locomotive's furnace is scanty, or if the draught is bad, less steam is made, and the power of the iron horse is diminished. If the food or fuel of the human machine is curtailed, or if the draught through the lungs is reduced, the capacity of the body for the evolution of power will be lessened. Muscular power, nervous power, and mental power too, are generated, chemically speaking, by the union of the food we have eaten and assimilated, and the oxygen of the air we breathe.

As a locomotive speeds on its way, puffs of steam and smoke escape from it every moment, evidence of the change of matter going on in the furnace, and of the power produced and consumed. These statements are true of the human body; every breath we exhale is composed almost entirely of the vapor of water, vaporized by internal heat and carbonic acid gas, a product of the combustion going on inside the body, by which the power we have is generated.

Still, although there are many points of resemblance between the way in which power is produced in the human body and in the steam engine, there are also some important points in which they differ. The fuel of the locomotive is simply cast into the fire and burned ; but the fuel of our bodies must undergo elaborate processes of organization. The food must become part and parcel of ourselves to a great extent, before it can be used in producing power. For instance, when a muscle contracts a portion of its substance is used up in generating whatever power is evolved. The used up part unites with oxygen.

The same changes occur in the brain and nervous

system, when these are called into physiological activity. Therefore, persons whose capacity to breathe is below the full requirements of the system are not capable of exerting their full quota of any kind of power, neither muscular, nervous nor intellectual. Therefore, if any one laboring under consumption or other chronic disease, desires more muscular strength, larger powers of endurance, increased capacity to be and to do what he will, who wishes to have the best chance there is to recover his health and to get rid of the load of weariness that unfits him for the active duties of life, and destroys so much of its value, let him get his lungs into as good working order as possible, and have his breathing capacity brought up, approximately, at least, to the normal standard, and he will be surprised and gratified to find how rapidly his strength will increase and his enjoyment of life be enhanced.

But the capacity to get into the lungs enough fresh air at every breath is necessary, not only to the evolution of power ; it is also essential to the perfection of all the vital processes by which life is

continued. If the lungs fail to do their work properly, the food cannot be thoroughly digested, it
cannot be converted into blood of the best quality.
The wonderful transformation by which the blood
becomes flesh, nerve, bone and tendon must necessarily be a partial failure. And to crown the whole,
the processes by which the system is relieved of
waste matters cannot be satisfactorily accomplished.
It is no wonder that those who breathe too little
air at each breath frequently become consumptive.
When I think of the extreme importance of the
respiratory function, and that it is so frequently
defective, I am surprised the results are no worse
than they are. And when we consider that even
the small quantity of air many persons do breathe,
is often far from being pure, it is a signal proof of
the great capacity possessed by the human constitution to maintain life under very unfavorable
conditions.

IS CONSUMPTION CURABLE ?

That there are numerous recoveries from this

disease has been proved by autopsies of the bodies of persons who died of other diseases. There are frequently found in the lungs of such subjects old dried tubercles, and sometimes the scars of cavities from which masses of tubercular matter had been ejected, showing that the individual did at one period of his lifetime recover from what is usually called the second, or even the third stage. It is astonishing how large a part of the lungs may be rendered useless by disease, and yet the invalid both lives and enjoys life with what remains, provided that can be kept free from disease.

Tubercles and their consequences must always be unwelcome occupants of the living body; but if they do occur, it is better to have them in the lungs than in almost any other organ, because nature has generously furnished us with more breathing area than is absolutely needed to sustain life. Every sound pair of lungs has a reserve breathing power by which oxygen is furnished to the body in adequate quantity, when severe muscular exertion is demanded. A few dozen small tubercles may, and do often exist in a man's lungs without causing enough disturbance to attract attention, but the same amount existing under the mucus membrane lining the digestive apparatus would cause severe symptoms, while if they were situated in the brain or spinal cord, fatal results would promptly follow.

A moderate amount of tubercle in the lungs need not discourage any one, provided the constitutional symptoms are favorable. A single deposit of tubercular matter is never fatal, if it can be prevented increasing by fresh accessions from the blood.

This disease is popularly regarded as being uniformly fatal ; so it is after it has attained a certain point in its progress. But lung disease is not commonly regarded as being consumption at all until it has arrived at the incurable stage. And if an individual recovers from an attack of ill-health during which he manifested the early symptoms of the disease under consideration, he and his friends are very apt to conclude he never had any consumptive tendency at all. But the fact, however, is that at a given point in the early history of two cases of lung disease, one of which ended disastrously, while the other recovered ; the disorder was identical in both, the only difference being (and a very material difference it is) that while the former went from bad to worse, the latter took a favorable turn.

The obvious lesson conveyed by these facts is that the earlier the most effective known treatment is adopted, the greater are the chances of a cure.

THE CURABLE AND INCURABLE VARIETIES OF CONSUMPTION.

There are two kinds of phthisis, the acute and the chronic. The acute variety runs its course in from three to twelve weeks. The chronic usually lasts from two to four years, and may continue five, ten, or even twenty years. Acute consumption is uniformly fatal, but, as I have stated, the chronic variety is often amenable to effective and timely treatment. The longer a case is in developing and the slower its progress, the greater is the probability of recovery. That is to say, if a man has much tubercular matter rapidly deposited in his lungs, which soon begins to soften and break down, this being accompanied by local inflammation and notable constitutional disturbance, his case is very apt to terminate unfavorably. But if his disease progresses slowly, although his lungs are locally injured, if he has not lost much flesh, and his temperature is still normal, he stands a good chance of recovery, if under effective treatment.

ILLUSTRATIVE CASES.

CASE 1.

Mrs. G. consulted me for a form of indigestion from which she had long suffered. While examining her case, she mentioned incidentally that she had a slight cough for the past month, but evidently felt no alarm about it. This directed my attention to her lungs, when, to my surprise, I found the apex of the left lung the seat of tubercular deposit, already undergoing softening. Believing that the unwelcome discovery would have a very depressing effect on this lady, I said nothing to her about it, but stated the whole truth to her husband.

As diseased lungs cannot be cured as long as the stomach does its work badly, that organ was first corrected by the use of proper remedies. This done, she took cod liver oil and iron, with a course of special training for six weeks. The cough ceased before the end of a month, and at the completion of the treatment her physical strength had greatly increased. The only evidence that the apex of her lung had ever been diseased was a harsh sound caused by the air rushing through the healed lung tissue. Several years after, I heard she was travelling extensively for pleasure in the Western States. The above is a case in which active lung disease was discovered, I may say, accidentally, and effective treatment promptly employed, by which a cure was speedily attained. I am convinced that if this lady had only taken the usual remedies, special training being omitted, her disease would have progressed practically unchecked.

CASE 2.

Mr. C., a gentleman in his 64th year, had been suffering from senile phthisis for over three years, during which time he had been under the treatment of physicians of the highest reputation for lengthened periods. For some time before I saw him, he had concluded that the healing art could do little for him, and he only consulted his family physician for the relief of annoying dyspeptic symptoms. His wife desired him to consult me. For a time he refused, declaring that any disease the noted medical men, under whose care he had been, were unable to help, could not be successfully treated. However, to get rid of the lady's importunities, he finally came to see what I had to say about his case.

I found he did the usual amount of coughing daily, had slowly lost flesh, his appetite was capricious, and he often spent wakeful nights ; but the difficulty that oppressed him most was severe shortness of breath. If he ascended a slight rise of ground, or quickened his pace above a very slow walk, he was soon obliged to stop, breathless ; and even when at rest his respiration was attended by discomfort. I found that although both his lungs were free from active disease, except a small part of the left lung, yet the inspired air penetrated into the air vesicles very imperfectly. The walls of his chest were unusually rigid ; taking a breath as full as he could, expanded his chest less than one inch in circumference. I told him he was going from bad to worse physically, not because he had extensive and active lung disease, but because his breathing power was steadily diminishing on account of increasing rigidity of the chest walls and stiffening of the whole structure of the lungs. I stated that if the size of his chest was enlarged the semi-collapsed lungs would spontaneously expand to fill the increased space thus provided for them. The inspired air would then readily penetrate to every part of the lungs ; moreover, if the

rigidity of the walls of his chest was replaced by
something like the natural elasticity, his breathing
would become habitually easy, and his wind during
exercise be greatly lengthened ; the reactions
occurring in the pulmonary air vesicles between the
breathed air and the blood would be thus promoted.
As the direct result of these changes, his appetite
would improve, his blood circulate better, the
disorganization going on in a portion of his left
lung would be checked, his strength improve, and
he would feel like a new man. And I added that
I thought every one of these desirable results could
be readily secured, mainly by a course of special
training adapted to his condition. The idea that his
breathing capacity could be increased was to him
most welcome ; and the whole theory I unfolded
of cases such as his, was so reasonable and so well
supported by his personal experience, that he at
once decided to place himself under my care. He
came regularly for twenty days. At the end of
this time he was so much better in every respect
that he decided to take a trip to the far West to

transact some very important business which had been neglected because he had hitherto felt utterly unable to undertake the journey. He was absent three weeks. After returning home he came under my care for another twenty days special training. The whole treatment so greatly improved his health that he went to Europe, where he travelled for several years, enjoying life, until his death, which occurred in Paris, in his 71st year, of a few days' illness of acute pneumonia.

CASE 3.

Mrs. H., a lady fifty years of age, had been in failing health for about three years before I saw her professionally. Her family physician, an eminent medical man, having failed to restore her health, advised travelling in Europe. She did so for a year, and while there tried the curative efficacy of the climate of Southern France, together with the treatment of the most celebrated European physicians, but came home much worse. Several months after her return she was induced to consult me by a gentleman who was aware of the valuable results I had obtained in another case then under my care. After examining her, I stated that there was no possibility of effecting anything like a cure, but I thought that relief would be obtained if her breathing capacity could be increased. I was led to this conclusion because her case resembled that of Mr. C., only farther advanced. In a few days her respiration was much easier, and after three weeks' treatment she became strong enough to ride out every suitable day, although she had been previously so ill that she never expected to be able to leave her house alive. She continued to improve pretty steadily during the three months of treatment, and held the ground gained for more than eighteen months. But the disease with which she had so long struggled again became active, and she passed away about two years after I first saw her.

CASE 4.

Miss W., a lady about twenty-four years of age, had been ill over one year before I saw her. She presented the usual symptoms of developed phthisis, the most noted of which was pulmonary hemorrhage. She had raised from a few drops to a teaspoonful of blood almost every day for the previous twelve months. Every case has its peculiar and leading indication. In this it was congestion of the lungs, indicated by two prominent symptoms. The blood circulated irregularly, it left the surface of the body and the limbs and stagnated in the lungs, causing coldness of the skin and extremities, with escape of blood from the bronchial tubes. To relieve the lungs of the surplus blood by which they are loaded is an essential part of the treatment of such a case. Without this there can be no cure. Remedies of various kinds, doubtless, serve very important purposes ; but I know of no medicine, nor combination of medicines, that has any power to regulate disordered blood circulation. By special curative training, however, the currents of the vital fluids are directed at will with almost the same facility with which we solve any problem in the exact sciences.

Before Miss W. had been under treatment two weeks, this end was so far attained in her case that the hemorrhage from her lungs ceased entirely. Her skin and limbs now became habitually warm by the better blood circulation flowing through these parts. At the end of about three months her disease seemed to be quite subdued, and treatment was therefore discontinued. I heard little of her for nearly eight years, when I met her accidentally in the street. She informed me that since her recovery she never raised any blood, nor had her old disease given her any trouble. She had caught cold several times during the winter season, as stronger persons are liable to do, but the resulting coughs had always ceased under the use of simple remedies. At my request she allowed me to examine her lungs, and I found them in better condition than when she ceased treatment eight years before.

CASE 5.

Four years before I met Mr. D. he had been an invalid. A large part of that time he had been searching for health. One season had been spent among the Rocky Mountains, another in Southern California, and the last in Aiken, South Carolina. He had also consulted the most noted medical experts in diseases of the lungs, all of whom agreed that he had fibrous phthisis, a somewhat rare form of the disease. But the only satisfaction he ever received was a scientific description of the pathology of his case, accompanied with an elaborate and accurate diagnosis. None of the remedies prescribed ever did him any good, and in some instances they were distinctly injurious. When he found that the most favored climates on this continent, added to the best medical treatment of which he knew anything, failed to materially check his disease, he settled down to the conviction that there was not even temporary help for him. At the time I saw him his breathing capacity had so diminished, that simply walking up two or three steps of a stair would cause breathlessness.

Breathing is, as I have stated, in the first instance, a purely mechanical process ; the respiratory muscles enlarge the size of the chest, when the elastic lungs swell out to fill up the increased space, and air rushes into the bronchial tubes and air vesicles. But my suffering patient's lungs had become abnormally solid and inelastic, and the walls of his chest were nearly motionless, even when he tried his best to take a long breath. In the fibrous variety of consumption a constant contraction of the lungs and of the cavity of the thorax goes on, and of course the capacity of the lungs to receive the breath of life is reduced in the same ratio. These sufferers feel as if the chest was surrounded by an iron band, which is being slowly tightened by a screw. Under these circumstances, breathing is distressing and laborious.

An inspection of my patient's chest showed that the mechanical conditions demanded for carrying on the respiratory process were sadly defective. The very first thing to be done for such cases is to make the chest bigger, but it is usually the very last thing recommended. As the outside of the lungs must always be in contact with the inside of the chest walls when the latter are expanded, the former also grow larger to fill the space provided for them. The capacity of the lungs to receive more air at once increases, and the wind becomes longer immediately. All this must be accomplished for these breathless sufferers, without any exertion on their part ; muscular effort rapidly causes congestion of the lungs, one symptom of which is breathlessness. Therefore, the curative training to which they are subjected must be conducted with great care and be entirely of the passive variety. My theory of his case pleased this gentleman so much that he decided to begin treatment under my direction. Although, he said, sadly, your ideas are plausible, but I am afraid my case is beyond aid of any sort. And in one sense he was right ; a real cure could not be attained when his disease was so advanced that a fatal result in the near future was certain. If the medical men whom he had consulted in the early part of his illness, had, along with the tonics they gave him, suggested the line of treatment he finally received,

he would certainly never have been in the condition
in which I found him.   Toward the end of the third
week's treatment, he said, " I really think my breath
comes easier, and is longer ; and I know my cough
is less troublesome."   He continued to improve in
every respect, and when he went away at the end of
two months' treatment, I never saw a patient more
thankful for very great and totally unexpected
help.

CASE 6.

Mr. S., a gentleman forty-four years of age, of
fine physical development, presented himself for
treatment.   His appearance was that of a man in
fair health.   Yet a noted physician of Philadelphia
had informed him that the apex of the left lung
was the seat of a deposit of tubercular matter.
I examined him and confirmed the diagnosis.
The only prominent symptom of which he com-
plained, was a cough that expectorant mixtures
had failed to benefit.   His appetite was good, and
although not quite up to his usual weight, he had

not lately been losing flesh. He was by no means very ill, but was well aware that he had in him the beginnings of a disease, which, if not cured, would be very apt to kill him in two or three years. This is just the sort of a patient I like to see ; one who really has something wrong with him that requires skill to cure, and yet is not passed into a condition for which little or nothing can be done.

To dry up pulmonary tubercles, and reduce them to small, hard, inert nodules, is the best results looked for under the usual mode of treatment ; but when special curative training is added, the absorption of the whole tubercular deposit and the restoration of the lungs to a healthy condition, is the result usually attained in suitable cases, when treatment is commenced at a sufficiently early period in the progress of the disease. Before he went home, at the end of three months' treatment, his health was good, and the thorough healing of the diseased part of the left lung was readily proved by the disappearance of the physical signs indicating the presence of softening tubercles, and the reappearance of the signs showing that the formerly diseased lung tissue was healthy. At my request, he promised to report to his former medical adviser on his arrival in Philadelphia, request an examination, and state by what means the cure had been attained.

These few cases, among many that might be cited, are recorded to show that the results attained in the treatment of chronic pulmonary disease by the means here advocated, supplemented by the use of such remedies as may be indicated in each case, are greatly superior—with regard to the rapidity with which patients improve, the greater certainty of the curative effects, and the permanence of the results—to all other means combined when special curative training is omitted.

The foregoing cases are not detailed for the purpose of leading the reader to think that every apparently similar case will do as well under treatment. The curative results attained in different cases differ greatly. Sometimes patients expect little benefit, yet get wonderfully better ; others look for a cure and fall far short of that desirable end. Every case must stand by itself. An amount of disease that is readily removable in one individual, defies our efforts in another. The patient's constitutional stamina has much to do in determining the result.

THE CHIEF DIFFICULTY IN EFFECTING CURES.

Consumptives are naturally slow to admit, even to themselves, that their disease is serious, and they are still slower to adopt any treatment that does not receive the hearty endorsement of the family doctor. The vast majority of cases of phthisis assume the chronic, and, therefore, curable form at the beginning ; but in their progress they assume more and more the acute type. At first the respiratory capacity is nearly normal, the weight of the body little, if at all, diminished, the pulse regular, the bodily temperature unvarying. Yet at this point a critical examination of the lungs will rarely fail to show that they are already undergoing destructive changes, and in a few months the breathing becomes frequent and shallow, flesh is lost, the pulse rapid, and the blood acquires a fever heat in the afternoon. The progress of chronic consumption may be aptly compared to that of a solid body moving down an inclined plane. At first its motion is comparatively slow, and a moderate force will arrest it, but if it is allowed to proceed it soon acquires a momentum so great that it sweeps away every obstacle in its path. At this point all remedial measures necessarily fail. If medical men generally knew the value of the measures here advocated, when these are properly employed, and promptly added them to the valuable nutritive, tonic, and hygienic treatment now adopted by all educated physicians, as soon as cases of incipient consumption came under observation, comparatively few cases of chronic phthisis would proceed to a fatal issue : but special training is rarely thought of, until much treatment has been fruitlessly tried. Therefore, the good we do is attained under very unfavorable circumstances.

SHOWING WHY AND WHERE THE PREVALENT
TREATMENT IS DEFECTIVE.

The fundamental difficulties in the disease under
discussion are lack of power to assimilate nutritive
materials and reduced capacity to breathe, the
former defect depending largely on the latter. The
consumptive fails to make the food he eats, the water
he drinks, and the air he breathes, into good, rich,
generous blood, fitted to perfectly nourish his body
and sustain it in health and strength. If he could
do these, no deterioration would occur in his vital
fluids, nor could any inflammatory action or tubercular deposit occur in his lungs. His failure to
accomplish these essential physiological processes
properly is the prime cause of his disorder.

Ever since physicians abandoned the depletory
system of treatment, the theory that the phenomena of this disease are primarily due to
defective nutrition, has been gaining ground, and
it is the view now almost universally accepted by
the profession. Differences of opinion, however,

continue to exist concerning the exact nature of
the nutritive element the system fails to appro-
priate. Thus the slow loss of flesh leads many to
prescribe cod liver oil ; others, believing that consumption is due to a want of lime in the system,
vaunt as specifics the salts of that substance.
Dr. Churchill, of Paris, recommends phosphorus
in certain combinations, therefore the hypophos-
phites of lime and soda are extensively employed.
Dr. Dobell, of London, thinks the disease is caused
by lack of power to digest and assimilate solid
fats ; he therefore favors the internal use of puri-
fied tallow made into an emulsion as essential to
successful treatment. The list could be greatly
extended, but these are sufficient to indicate the
class of remedies generally used by the most repu-
table physicians in the treatment of this disease.
The human body requires a large variety of
nutritive materials to nourish it perfectly. For
instance, food must contain, in suitable proportion,
chlorides of soda and potassium, carbonates of
lime and soda, phosphates of lime, soda, magnesia

and potassa ; besides these proximate principles of Dr. Dalton, nitrogenized organic matters, such as the albumen of eggs and the constituents of lean meat, are demanded to build up the body. While food containing carbon is necessary for the production of animal heat, of this latter variety butter is a familiar example. Now, if the diet of a man does not contain these nutritive principles in due proportion, disease must soon be developed in him. Thus scurvy is believed to be caused by want of the salts of potassa contained in most fresh vegetables. And the valuable remedial effects of iron in many diseases prove that the presence of a due proportion of that metal in the blood is essential to health, its absence giving rise to the anemic condition.

Consumptives usually have too little fat on their bodies, therefore cod liver oil, pancreatic emulsions and other readily assimilable fats are good remedies. Their blood is undoubtedly more or less deficient in the chemical elements which that fluid should contain ; therefore, lime, phosphorus, iron, etc., would seem to be the remedies from which great

results might be expected. The treatment of
physicians who prescribe them is, therefore, theo-
retically correct, but practically of little value. It
is true as far as it goes, but being narrow and one-
sided, is fatally defective and therefore unsuccessful.
It is a move in the right direction, but farther
advance is necessary before solid success can be
achieved. Phthisis is truly characterized by defective nutrition, as before stated ; but the disease is
not often developed in this land of plenty for lack
of an abundance of nourishing food. The consumptive's blood has not deteriorated, nor his lungs
become diseased because he did not have plenty of
suitable food on his table, or, for that matter, in
his stomach. They did become diseased, because
having eaten the food, he was unable to properly
digest and assimilate it. To meet and correct the
depraved condition incident to this disease, many
staple articles of diet have been concentrated to a
fraction of their natural bulk with a view to increase
their nutritive power for this class of invalids.
Fatty matters, lime, phosphorus, iron and all the

other elements of nutrition exist in the phthisical sufferer's food in adequate quantities, but his vital chemistry cannot combine them into the forms needed for the vital purposes. There is no question but that it is a decided advantage to have the special elements in which the sick man's blood is defective present in his food in excess. This fact alone is a sufficient justification for the extensive use of the popular remedies for lung disease. But after this valuable advance in the right direction has been made, there remains to be corrected a very serious defect in the present treatment of consumption. It is not enough to supply a consumptive with all the materials he needs—nourishing food and nutritive remedies—then leave him to build himself up with these as best he may. IIis power to do this can and ought to be increased. His capacity to make richer blood, stronger muscle, and sounder lung tissue can and should be enlarged by stimulating his assimilative functions to the highest possible degree of activity. When these facts are fully recognized and acted on, the treatment of consumption will be placed on a more solid scientific basis than ever before, and the prognosis of this dreaded disease will be more hopeful.

Dr. Chambers, of London, says, "It is truly by the aid of the digestive viscera alone that consumption is curable. The chest is the battle ground of past conflict, the stomach the ripening ground for new levies of life. Your aim should be to get the greatest possible amount of albuminous food fully digested and applied to the purpose of the renewal of the body, at the same time that the renewing agencies are brought to their highest state of efficiency. In this way a healthy cell renewal takes the place of that morbid cell renewal that appears in the shape of tubercular matter."

The views of this distinguished medical teacher are right as far as they go. To increase the appetite of a consumptive to a healthy standard, and enable him to make his food into good blood is to cure him. But the amount of food a man can digest and assimilate bears a very close relation to his respiratory capacity, and the rapidity with which tissue change goes on in his body. If an abundance of oxygen finds access to a man's system, it completely reduces the bodily waste to those forms by which it is fitted to be readily and perfectly eliminated from the body. Thus two atoms of oxygen unite with a dying particle of organic carbon. These united in this proportion form poisonous carbonic acid gas, which, if it accumulates in the blood, rapidly causes death ; but having the gaseous form, it escapes through the lungs as quickly as it is formed. In the same way oxygen unites with decaying particles of nitrogenized tissue, and from their chemical union poisonous compounds arise, but being soluble in water they are promptly eliminated by the kidneys.

In this way oxygen, by removing waste matter from the body, makes way for new material. The intricate nutritive operations inside our bodies, in one respect at least, resemble mercantile transactions in the open market ; the demand in both cases regulates the supply. If the former is brisk, the latter will be abundant. It is precisely at this

point the current treatment of consumption fails
The urgent necessity of a larger supply of nutritive
materials to the sick man's vital manufactory is
perceived, and supplies are therefore poured in ; but
as no effective attempt is made to create a genuine
demand, his vital market is soon glutted. Little
or no account is made of the fact, that if an individual's respiratory capacity is reduced, the removal
of waste matter from his body will be reduced in
the same ratio, his appetite immediately and necessarily falls off, and his physical nutrition becomes
disordered.   But if we increase his breathing power,
an increased demand is set up in his system that
very hour, his nutrition improves and his stomach
becomes clamorous for an augmented allowance.
Thus we see it is impossible to permanently
improve a man's appetite so long as his capacity to
breathe steadily declines.   But if we can augment
this, the oxygen thus introduced into the system
will make change of tissue active, and the appetite
will take care of itself.

THE MECHANISM OF BREATHING.

Inspiration and expiration, or the motions by which air passes into, and is expelled from the lungs, are caused by the alternate expansion and contraction of the chest, produced mainly by the power of the diaphragm and the muscles situated upon and between the ribs. There are, however, besides these, the auxilliary respiratory muscles that come into play, whenever forced breathing is performed. While the breath is passing into the lungs the muscular floor of the chest descends, and the other respiratory muscles draw the ribs upward and outward. By these two actions the size of the chest is increased in every direction. While the air passes from the lungs the diaphragm rises, the ribs fall, and the chest contracts. When the breath is inhaled the whole front of the body gently swells outward ; when the breath is exhaled it falls inward. When the chest is enlarged by muscular action, air rushes into the lungs as into a pair of bellows when the sides are drawn apart.

The air is again expelled from the lungs by the contraction of the chest, as from the bellows, by the closing of its sides.

When this vital gas is once in the system, it gives rise to many wonderful chemical and vital operations ; but the act of breathing by which it is introduced into the lungs is purely mechanical, consisting simply in enlarging the capacity of the chest by muscular action, when air rushes in to fill the vacuum, as before stated.

But consumptives lack to a great degree the mechanical conditions requisite to carry on the respiratory process perfectly. Their chests are usually narrow and flat, not large, round and roomy ; the walls are preternaturally stiff and unyielding, not elastic, playing out and in freely at each breath. The lungs themselves also lose their resiliency more or less, and some parts become condensed almost as solid as liver. The efficiency of the whole lung is thus impaired, and the functions of some parts entirely destroyed, not only by the deposit of foreign matter, but also because too little air enters them to preserve their normal inflated condition. If we cease to breathe with any portion of our lungs, the parts not used will soon be unfit for use.

THE EFFECT OF EXERCISE ON HEALTHY PERSONS.

The beneficial influence of exercise on health is universally admitted. Infinite wisdom has inseparably connected physical toil with man's earthly well-being. Early in the history of the race, the Creator said, " By the sweat of thy face shalt thou eat bread." The riches of the wealthy do not exempt them from the common lot. This truth is tersely embodied in the proverb, "The poor man must work to find food for his stomach ; the rich man must work to find a stomach for his food." The superior health enjoyed by those accustomed to active out-door life, often, in spite of many injurious habits, conclusively proves the value of exercise for the preservation of physical vigor. The moderate toil of a healthy man deepens his breathing, strengthens his muscles, sweetens his rest, purifies his blood and secures its vigorous circulation—in short, it keeps up in his system all those chemical and vital changes that are essential to health and strength.

THE EFFECTS OF EXERCISE ON INVALIDS.

The voluntary muscles are so called because their action is under the control of the will. They are made up of bundles of fibres bound together by connective tissue. Each of these is capable of being separated into smaller bundles until only a single fibre can be removed. This fibre is, however, capable of being divided into primary fibrillae, each of which is seen under the microscope to be transversely marked by fine lines which causes them to present the appearance of a string of minute beads. The lines are produced by alternate light and dark muscle cells composing the primitive fasciculi. When a muscle contracts it becomes thicker and shorter ; this change in its shape is caused by an analogous shortening and widening of the muscle cells in which the inherent contractility of muscular structure resides. When a voluntary muscle is used too little or not used at all, its structure undergoes a diseased change—fatty degeneration.

When it is normally used it undergoes healthful destructive change. The primary effect of muscular action, which results in the bodily movements, whether of work or play, popularly called exercise, is always to cause a loss of substance in the acting muscles. Muscular contraction occurs in obedience to the mandates of the will transmitted through the nervous system, involving expenditure of nervous power. The loss of nervous energy is always much greater in rapid than in slow muscular motions. Repair of the retrogressive change that takes place in muscles is effected under the influence of the nervous system. When the exercise of an individual's muscles has been of the right kind and amount, and the expenditure of nervous power not greater than he can afford, the loss of substance suffered by the acting muscles is quickly replaced by new material. Under such circumstances the size and strength of muscles progressively increase

up to a certain point.   But if the exercise has been
of an improper kind, or excessive in amount, the
loss of nervous vigor will also be in the same ratio.
Then the renewal of the wasted muscular structure
will either not take place at all, or it must do so
slowly and imperfectly.

Therefore, when we prescribe exercise for consumptive and other chronic invalids, we should
instruct them as to the kind, quality and amount
adapted to their condition.   We must not forget
that exercise has its laws that cannot be disre-
garded with impunity, at least by sick people.   It
would be almost as wise to expect good results by
ordering a sick man in need of medicine to enter a
drug shop and take a dose of the first mixture he
saw, as to advise sick folks indiscriminately to exercise as a means of cure.   The truth is, that
although muscular exercise, as we have seen, is
admirably adapted to promote the vigor of fairly
healthy persons, it is ruinous to invalids after they
have sunk below a certain point.   There are multi-
tudes of delicate people who feel deeply their need

of the health-giving influences of exercise, who are bitterly conscious of being injured every time they indulge therein. For instance, a consumptive is advised to exercise ; he accordingly makes what is to him a dangerous experiment, and too often, instead of being benefited, his condition is made worse. The explanation of this is, that to such persons general muscular action, when sufficiently vigorous to be productive of a curative amount of tissue renewal, is unduly exhaustive of nervous power.

Exercise affords weak invalids, laboring under the disease we are considering, little or none of the invigorating effects that flow to the healthy worker from his labor. Their meagre nervous energies are rapidly exhausted. Exercise prostrates them instead of inducing an agreeable sense of fatigue soon relieved by rest. It weakens rather than strengthens, because it occasions a waste of nerve and muscle that they have not the vitality to repair. When muscular action occasions exhaustion, its effects are always injurious in consumption and other chronic disorders. This point is reached even before the invalid is aware he is moderately tired. Yet although such sufferers cannot profitably indulge in vigorous exercise, they are urgently in need of the bodily renewal to be secured thereby. They need the deep breathing, strong muscles, sound sleep, keen appetite, pure blood, vigorous circulation, and quiet nerves; in short, they need the perfect nutrition, the rapid bodily renewal which is the constant attendant of well-regulated muscular activity.

The course of the blood through the body has been described as the lesser or pulmonary circulation, and the greater or systemic circulation. The current of the first proceeds from the right ventricle of the heart, through the pulmonary arteries into the lungs, in which the blood gives off carbonic acid and receives a supply of oxygen; after which it flows through the pulmonary veins to the left side of the heart, from which it passes by the general arterial and capillary systems to all parts of the body. While performing the latter circuit, the vital fluid absorbs carbonic acid and surrenders oxygen.

When the muscles are in vigorous action, the necessity for an abundant supply of oxygen to them and the rapid removal of the carbonic acid imposes for the time being increased work on the lungs.

The normal number of respirations in a healthy adult is about eighteen per minute, but violent exercise will increase these to thirty-five or forty, in the same time. So long as the power of maintaining deep and rapid breathing continues, oxygen is introduced into, and carbonic acid is removed from, the blood with sufficient rapidity ; but when this power fails, the supply of oxygen diminishes, and carbonic acid gas accumulates in the blood, causing "breathlessness."

For instance, the distress experienced by oarsmen after a severely contested but short race, is due to the breathless condition ; they suffer comparatively little from fatigue ; the muscular strength of athletic men cannot be so rapidly exhausted. The breathless condition is due to several other causes.

First, the increased action of the heart sends so large a volume of blood to the lungs that the vessels through which it must pass are unable to transmit it, even if the blood was not unduly charged with carbonic acid gas. Second, but as it is so charged, its affinity for the walls of the vessels is diminished, and as a necessary consequence its circulating power is reduced in the same proportion. The result of all this is that the blood passses through the lungs less rapidly than through the rest of the body. Great congestion of these organs is thus caused.

If this is the effect of active exercise on the most athletic men, it is easy to understand how breathlessness is so readily produced in those laboring under chronic disorders of the lungs. Their rigid chest walls and inelastic lung tissue, partly obstructed by deposits of foreign matter, are unable to respond to the increased action of the heart that occurs in such persons during very moderate exercise.

Therefore, if an individual in whom tubercular

lung disease is being developed, tries to secure the physical renewal by active exercise that the necessities of his case demand, and without which a cure can scarcely be expected, it will still farther shorten his impaired breathing, increase the action of the heart, and send the blood to his congested lungs with so great a rush that it often finds vent by bursting from these weakened organs.

THAT WHICH CONSUMPTIVES MOST URGENTLY NEED.

The life functions move, as it were, in a circle; all the vital and chemical changes by which life is manifested are mutually interdependent. Again, every grave disease has its prominent lesion; in one it is the digestive, in another it is the nervous; and in a third it is the respiratory system that is mainly at fault. Of course, disorder occurring in any of these unfavorably affects all the rest. In the disease under consideration there is no question but it is failure of the respiratory capacity that offers the prominent impediment to recovery. If this state-

ment is true, and I think it cannot be denied, it
necessarily follows that successful treatment must
consist, at least in part, of measures adapted to
increase the respiratory power. There is, doubtless, an inherited predisposition that often has
something to do in the development of consumption ; but I feel confident that this congenital
tendency would be developed less frequently than
it is, if the breathing powers were carefully and
scientifically cultivated.

The inhalation of pure oxygen gas has been
recommended to supplement the impaired breathing
capacity of consumptives ; but this treatment has
never been productive of valuable results. The
reason is obvious. The Creator made one hundred
parts of atmospheric air to consist of about twenty
one parts of oxygen and seventy-nine parts of nitrogen. These proportions cannot be altered so as
to make the mixture more suitable for breathing
either during health or disease.

Consumptives do not need condensed air, rarefied
air, or an artificial mixture of respirable gases.

They need pure, fresh air, and power to breathe enough of it.

In order to secure this the capacity of the chest must be enlarged, and the motions of its walls restored, so that both the chest and the lungs will freely expand with every breath. The requisite quantity of air will thus be secured without perceptible effort. The respiratory muscles must be invigorated, and the capillary blood circulation rendered as nearly perfect as possible. In proportion as these indications are acomplished, the breathing becomes calmer and deeper, even when indulging in exercise. The blood is more abundantly supplied with oxygen ; the reviving life forces soon feel the influence of the vital gas, their natural stimulus and the symptoms of disease give place to the evidences of returning health in proportion to the power of the patient's vital forces to respond to effective treatment.

WHAT SPECIAL TRAINING CAN DO.

Training, says Dr. Ralfe, is the art which aims at bringing the body into the most perfect condition of health, making muscular action more vigorous and enduring, and increasing the breathing power. The late Dr. Parkes says : " Training is simply another word for healthy and vigorous living. The objects of training are obtained by the employment of a system of diet, regular and systematic exercise, and a scrupulous attention to matters of personal hygiene." Strong men in sound health are always required to undergo training before engaging in athletic contests to bring up their health to the highest attainable point.

Now, the very great efficacy of exercise as a means of promoting health as well as the harm often resulting from its injudicious use in certain diseased conditions being admitted, it becomes a question of great practical importance in a medical point of view, how can we, in any given case of disease, secure the good results exercise is so capable

of producing, and avoid the bad ? Although this
question has puzzled many otherwise learned and
able physicians until the present day, it has been
solved in a thoroughly practical manner. Before a
physician can prescribe a medicine successfully, he
must have an accurate knowledge of its toxic and
therapeutic effects on the human body, and a sound
knowledge of the disease under which his patient is
laboring. These statements are true of the healing
use of exercise ; but that it may become curative
the prescriber must have an intimate knowledge of
the good or evil it can do, and he must labor under
no error concerning the physical capacity of the
patient or of the nature of the disease for which the
remedy is recommended. Therefore, in order that
exercise may become curative, it must be prescribed
and applied in accordance with the indications.
The quantity and quality must be scientifically
adapted to the therapeutic needs of the patient.
Under these circumstances it becomes a medical
resource of undoubted efficacy. The principles of
exercise are thus reduced to perfect order, its great

therapeutic value shown when these are properly understood and practiced ; and light is shed on a hitherto sadly neglected but valuable branch of the healing art.

By special training, the invalid gains all the good effects that flow to the strongest persons from general exercise, without subjecting himself to the injury it so often inflicts on weak persons. By its resources the chest can be expanded, the play of its walls increased, the surplus blood by which the lungs are loaded can be drawn away and distributed equally throughout the body ; all the muscles, group by group, can be brought into vigorous localized action, the appetite improved, the digestion and assimilation of food and nutritive remedies promoted, at the same time the breathing is deepened and prolonged, the frequency of the pulse reduced and the nervous energies husbanded. The weakest patient under special training is thus unconscious of fatigue; on the contrary, weariness is removed, and his strength steadily increases.

The intelligent reader will readily perceive that

treatment capable of attaining these objects must be highly curative in the disease under consideration, as well as in other chronic disorders requiring tonic treatment.

EFFECTS OF SPECIAL TRAINING ON THE BREATHING ORGANS.

By means of special training, all the contracted respiratory muscles can be gently but effectually stretched, the circulation in them improved, and their strength increased, rigidity of the thoracic walls overcome, and the chest vigorously and safely expanded. The air is thus made to penetrate into and inflate collapsed portions of the lungs, and dislodge matters with which such parts are obstructed. I may here remark, that no attempt should be made to expand the chests of persons suffering from consumption until after the blood circulation has been regulated.

This disease is usually limited to one part of the chest, at least during the early stages, while a cure is still possible. All the respiratory muscles are

stiff and weak, but the muscles covering the diseased side are always the stiffest and weakest. The
walls of the chest are, as we have said, contracted,
but the part covering the diseased side is always
more rigid and inelastic than that covering the
healthy lung. This is so palpable that an experienced physician need find little difficulty in point-
ing out the diseased part by these indications alone.
Special training, locally applied, is therefore required to overcome these obstacles to effective
respiration. By the means indicated I have seen
the muscles covering the chest and those between
the ribs become softer and greatly increase in
strength in a few weeks, the chest walls regain
their elasticity to a great degree and the flattened
side over the diseased lung becomes almost as full
as that on the sound side.

THE INCREASED AIR-SPACE USUALLY OBTAINED IN THE LUNGS BY SPECIAL TRAINING.

A healthy adult breathes about eighteen times per
minute, one thousand and eighty times per hour,

and nearly twenty-six thousand times in twenty-four hours. The quantity of air changed in the lungs in each act of ordinary breathing is from thirty to thirty-five cubic inches. Now we have seen that the quantity of air flowing into and from the lungs of persons in whom incipient consumption is being developed is always below the normal amount. To prevent this from steadily diminishing, or better still, to increase it to the healthy standard, is to them a matter of great moment. This is precisely what special training can do. It is not at all uncommon to succeed in augmenting the breathing capacity of such patients two, three, or even four inches at each ordinary respiration by skillfully applied training. But if we suppose the capacity of a patient's lung are increased by only one and a half cubic inches at each breath, he would then inhale fully twenty-seven cubic inches more air per minute, sixteen hundred and twenty more per hour, or nearly twenty-three cubic feet extra each twenty-four hours, an amount that would in many cases

stay the progress of the disease, and eventually lead
to its removal.

SHOWING WHY TRAINING TO BE CURATIVE IN THIS DISEASE MUST BE SPECIAL.

When a skilled trainer undertakes to develop the
physical powers of a crew of oarsmen, the material
on which he operates is carefully selected specimens
of humanity—young men who have already distinguished themselves for strength and powers of
endurance; his object in subjecting them to physical
discipline is simply to increase their health and
vigor to the highest point. This is precisely
what is wanted by a man who feels that his system
is in danger of yielding to a fatal disease. In
attaining this object for a boating crew of young
athletes, they are subjected to the same diet and
training, both of kind, quality and amount; but
when training is to be adopted as a part of the
treatment of consumptives this sort of generaliza-
tion is impossible. The training to which they
should be subjected must be special, in the most

restricted sense of that term. For instance, an individual affected by the disease under consideration, must have his chest expanded; but the amount of expansion to which it should be subjected must be determined for each case, from day to day, and sometimes as frequently varied according to the effects of the previous treatment. In many cases the dilatation to be perfectly safe must be slower and more gradual ; in others the rigidity of the chest walls is quickly and safely overcome.

All his muscles must be vigorously exercised ; but the training must be of such a character that while he is taking it his breathing shall not be hurried, the frequency of the heart's action increased, nor shall any more blood be sent to the lungs than when he is at perfect repose.

Training is requisite to promote the rapid renewal of the body—to break down old well-worn matter, and thus make room for new material, but no greater quantity of dying tissue should be hastened to its end each day than the oxygen being

introduced into the body through the impaired
lungs is capable of reducing perfectly.

Every act, whether of body or mind, involves loss
of nervous power ; therefore, training cannot be
conducted without more or less expense in that
direction, but special curative training must be so
conducted that the expenditure of the patient's
nervous energy shall be kept at the lowest possible
point. The valuable physical results that accrue
to a well man from gymnastics or other exercise
must be secured to the sick man, while the waste
of nervous energy must be sustained by the
operator who applies the special training. Otherwise, bodily renewal, to a curative extent, cannot
take place, as the latter occurs normally only when
the former is abundant. These are the prominent
points in which special differs from general training.

PRACTICE OF SPECIAL TRAINING.

In obedience to physiological law an increased
afflux of blood flows to organs in vigorous action,

and the system responds much more readily to the local demand if all the body, except the acting parts, is at rest. For instance, the head of a thinker sitting at his desk in deep study, soon grows warmer and his feet colder ; the temperature of the hand with which he writes will also be higher than that of the other resting passively before him. Through the operation of this law, carried effectively into practice by the use of active and passive localized muscular action, the pulmonary congestion which is a source of much distress to consumptives, and a prominent impediment to recovery, is dissipated ; the surplus blood oppressing the lungs and opposing their action is distributed to the skin and extremities where it is urgently needed. The muscles, with few exceptions, act in groups ; thus one group flexes the leg, another extends it. One series of muscles bends the arms ; another straightens them. By bringing all the various groups of muscles successfully into local action, with intervals of rest between each motion, while the remainder of the body is passive, the above described results are, in favorable cases, quickly and thoroughly attained.

Muscular contraction occurs because muscular fibre under suitable stimulus shortens, the diminution of length is accompanied by a corresponding increase of diameter. But this action takes place at the same instant, only in a few of the fibres composing a muscle, and in a portion of their length ; the contraction does not take place in the whole muscle suddenly, but, like a wave, it is gradually propagated along the entire length of the muscle, and through its thickness, one part relaxing as another contracts. The varying power exhibited by contracting muscle depends less on the energy with which each muscle cell contracts as on the number brought into action at one time. Again, the greater the power a muscle has to overcome, the more numerous will be the fibres contracting at the same instant, and the nutrition of the muscle will be improved in the same ratio. Therefore, in the use of special training, the muscular action elicited should be opposed by the hand of a skilled operator with a degree of power adapted to develop the strength and nutrition of the whole body. The power of a muscle is at minimum at the beginning and end of its contraction ; it is greatest in the middle of the act ; it rises and falls like a swell of music. Therefore, in order to secure the best results, the power opposed should increase and diminish correspondingly.

The expenditure of nervous power is relatively much greater when muscular action is rapid then when it is slow ; a man is less likely to be exhausted by a walk of ten miles than by a forced run of a single mile. The highest curative results are secured in special training only by exciting the maximum muscular action of which the patient's condition will admit at the least possible expenditure of nervous power. Therefore, the power opposed to muscles under training should be so regulated that while their contraction is not prevented, it shall be made to occur slowly.

The nutritive elements of the blood pass through

the walls of the capillaries to nourish the body only while the vital fluid is flowing through these vessels. Pressure on the walls of the capillary blood vessels has the effect of causing the blood to give up more rapidly and fully to the needy tissues the nutritive matters it contains. This physiological fact obviously has a very wide and important application in the cure of every chronic disease characterized by impaired nutrition. Therefore, the circulation through the muscles and the nutrition of the body generally should be stimulated by intermitting pressure with rotary kneadings of the soft parts, alternated with the more active special training already described.

Moreover, the use of special training modifies the effect of medicines in a very important degree. Patients who, before resorting to the treatment here advocated, took cod liver oil, pancreatic emulsions, cream, etc., in large doses, with little apparent benefit, often gain flesh notably after special training is added. Others, whose anemic condition indicated the use of iron, yet were not appreciably helped thereby, quickly show by the improved color of the mucus membranes and skin that the metal was being absorbed into, and was improving the quality of the blood under the influence of special training. Others, who were obliged to depend on frequent doses of whiskey or rock and rye, to overcome temporarily the languor by which they are oppressed, are, after a time, enabled to dispense with artificial stimulants by the natural stimulus of improved health.

Again, when special training is adopted, the medicines formerly used with more or less benefit not unfrequently require to be replaced by other remedies better adapted to enhance the curative effect of the new treatment.

THE EFFECTS OF SPECIAL TRAINING ON THE CIRCULATION OF THE BLOOD.

In attempting the cure of consumption, one of most imporant indications is to draw away from the lungs the surplus blood with which they are congested, and to distribute it equally throughout

the body.   This congestion occurs both in the
capillaries which carry the venous blood through
the lungs for purification, and in the vessels bringing blood for the nutrition of the lungs themselves.
The dull pains felt by those suffering from chronic
lung disease are largely due to stagnation of blood
in the lungs.   The correction of this difficulty
always affords the sufferer very great relief.

When special training is used with tact and
judgment, the following effects may be looked for
in a few days :

All the blood in the capillaries is pressed into the
minutest ramifications of these vessels in greatly
increased quantities, and gently urged onward into
the veins, through which it must pass to the lungs
for purification.   The emptied capillaries are
quickly refilled with fresh blood from the arteries.
These latter vessels are also made to bring a larger
supply of arterial blood charged with nutritive
matter, which is given up to the needy parts while
the blood is passing through the capillaries.   The
veins (one of whose functions is to remove waste)

are stimulated to absorb this, and to dispose of it by the proper channels. The circulation of the blood is thus perfectly under the control of special training. The movement of the vital fluid through the body in favorable cases is made so perfect that the patient will feel the whole person, to the ends of the fingers, tingle with renewed life.

Congestion of the lungs is thus promptly relieved; the surplus blood, which was a short time ago oppressing the lungs and opposing their action, is now used to make the body glow with vital warmth.

The general nutrition of the body is also greatly improved, and a condition of both its solids and fluids is established directly opposed to that diseased change in the blood, which results in the deposit of tubercular matter in the lungs.

THE INFLUENCE OF SPECIAL TRAINING ON THE PRODUCTION OF VITAL HEAT.

Modern physiologists have conclusively proved that the evolution of animal heat arises by the slow and measured burning of the body itself, and the food, by the oxygen of the respired air. If a very combustible substance like alcohol is taken into the system, its combustion quickly liberates heat and produces a temporary feverishness. If an energetic supporter of combustion, like laughing gas, is breathed, heat is also rapidly evolved in the body. On the other hand, if the breathing goes on as usual, while the quantity of food is unduly reduced, the bodily weight diminishes.

Dr. Draper says, " As cold is felt from want of food, so also is it from want of air. In ascending high mountains the effect upon the system has been graphically described as cold to the marrow of bones." The explanation of this is clear. In the accustomed volume of air received at each inspiration the proportion of oxygen is less the higher we ascend into the rarefied atmosphere.

The fire inside the system that supplies the vital heat is governed by the same physical laws as the fire that heats our stoves. Diminish the food of the body or the fuel of the stove, or reduce the breathing capacity of the lungs, or the air draught at the stove door, and the result is practically the same—less heat.

We have seen that even in the earliest detectable stage of pulmonary disease the draught through the lungs to the vital fires is reduced, the change of tissue is too slow, and the appetite rarely good.

In view of the facts just stated, it is easy to understand why a man, laboring under these difficulties is not comfortably warm in ordinary winter weather, even when he has on all the clothing he can well carry about. By training, his breathing power can be increased, his tissues renewed, and his appetite healthfully stimulated—causes that always produce the proper effects, one of which is increased evolution of bodily heat.

REVIEW OF THE OBJECTIONS SOMETIMES URGED AGAINST SPECIAL CURATIVE TRAINING.

FIRST OBJECTION.—*As it is applied externally .it cannot reach any internal disease.*

This objection can only arise in the minds of those who are ignorant of the physiological effects of training, both general and local; yet as I have known it to be urged I shall briefly notice it.

Food applied to the mucus membrane lining the stomach has no nutritive effect until after it has been digested, and passed through the membrane into the blood. A dinner is, therefore, in the first place, really an external application, and if it does not happen to agree with the eater it may never be anything else, but be ejected by emesis. A dose of medicine or poison cannot exert either therapeutic or toxic effects, until after the drugs have been dissolved and absorbed into the blood. The above objection is therefore quite as applicable to medical treatment generally as to special training.

The general training undergone by oarsmen preparing for a race may seem to a superficial observer to be altogether external, yet by means of it their good health becomes better, showing that the vital processes have been thus very favorably influenced. But special is much more effective than general training, because it is adapted accurately to the peculiar needs of the individual. It truly places at our disposal means whereby the functions of life in the innermost recesses of the body may be aroused, stimulated and directed into those channels that most surely favor the recovery of health.

SECOND OBJECTION.—*Such treatment must be dangerous to persons who have a tendency to hemorrhage.*

An objection of this sort might be advanced against every sort of effective medical treatment. That which cannot do harm seldom has much healing virtue. There are few drugs of known therapeutic power which will not cause poisonous symptoms if given in large doses, or at the wrong time. In the same manner the measures here advocated are quite competent to injure a delicate invalid very seriously if not used with skill and judgment. In subjecting patients to special curative training, whether they have had pulmonary hemorrhage or not, all attempts to expand the chest should be deferred until the engorged condition of the lungs has been relieved, and the systemic blood circulation equalized. Then these operations may be resorted to, not only with perfect safety, but very great benefit. Special training, when used with discretion, never causes bleeding from the lungs, while, by the control it exercises over the circulation, it offers one of the most effective of all known means to prevent and arrest it.

THIRD OBJECTION.—*Physical training must be harsh and exhaustive, therefore it is unsuited to the treatment of delicate persons.*

This idea can only be entertained by those who are entirely unacquainted with the methods employed. Over-training is carefully avoided when preparing strong men for athletic contests, because of its disastrous effects on their health. For obvious reasons, much greater care and skill is necessary in the use of special training for curative purposes. Before beginning treatment the patient is carefully examined, not only by the approved methods of diagnosis practiced by educated medical men, but when conducting operations the physician subjects his patient to actual handling ; his sense of touch becomes educated, so that he is enabled by this means to form a more accurate estimate of the invalid's vital stamina than is possible by the ordinary means of diagnosis alone. This done, a course of training is selected, believed to be adapted to the individual seeker of health. Therefore, no greater error can be committed than to suppose that this mode of treatment is harsh, seeing it is suscep- tible of almost endless modifications, and can be adapted by an expert to the treatment of the most delicate bed-ridded women, whose muscular and nervous systems have been prostrated for years, rendering them altogether incapable of vigorous voluntary exercise.

Instead of being fatigued by special training adapted to his condition, an invalid feels decidedly refreshed and encouraged. His whole person glows with a genial warmth, the immediate result of equable blood circulation, the respiration becomes calmer and deeper, the pulse less frequent, more regular and strong. If pain exists, it is removed, or, at least, alleviated. Weariness is dissipated and the nervous system soothed.

THE ADVANTAGES OF SPECIAL TRAINING.—WHEN PROPERLY USED IT NEVER INJURES.

When a physician administers a drug, he is too often obliged to accept the risk of doing his patient some injury, that he may afterward accomplish for the sick one a greater good. This uncertainty is the necessary result of our yet imperfect knowledge of the physiological action of medicines on the infinite variety of constitution and idiosyncrasy presented by different individuals. But special training, so far as it goes, has no such drawback. With due care in its use the patient may be absolutely certain that no injury can be done. The impressions produced on the body by it are entirely in the direction of physiological growth and development. It enables the natural tendencies of the system toward health to act more efficiently. It directs the physical energies into those channels where they are most needed. It enables the system to develop and maintain its natural forces in greater amount, and the healing effects are produced without wasting the vital powers.

THE CURATIVE EFFECTS ARE QUICKLY APPARENT.

As the aim of special training is not merely to remove or change symptoms, but to eradicate the causes of disease, root and branch—in short, to effect as complete a renewal of life as possible—time is required; still a much shorter period is needed than might be expected, when the importance and permanence of the results are considered. The rate of improvement in different individuals varies considerably, being determined by the nature and stage of the disease, and the remaining constitutional stamina.

When patients are under ordinary treatment, without the measures here advocated, even in climates that are believed to be best suited to the cure of lung disease, they are happy if they experience a very moderate amount of improvement in three, or even six months. But when special training is added to the means usually employed, the results are altogether different. Persons of whom a favorable opinion is expressed will, with scarcely an exception, know themselves to be decidedly better in from ten to fifteen days ; an additional period of time is, of course, always required before the invalid attains all the benefit he is capable of receiving.

Confidence in any remedial measures a sick man may be undergoing greatly assists the physician's endeavors. If a seeker of health lacks faith in special training before coming under its influence, the encouraging accounts he receives from his fellow-patients, who have been under treatment a sufficient length of time to test its value, and the improvement he soon experiences himself, speedily fill him with hope and courage.

THE ACTION OF MEDICINES WITH SPECIAL TRAINING.

The natural tendency of the vital forces to right themselves is always the most important factor in every recovery from disease. Whatever questions may divide medical men they are all agreed that the true office of the physician is to assist nature as best he may. Now, in every important disease like that under discussion, nature's defences are weakened at several points, and if the physician is to accomplish all the good that can be done, he must not fail to fortify every dilapidated segment of the circle of life. The enemy cannot be kept out of the citadel by simply closing one open gate, although that may be the biggest, while he is allowed free entrance by several other openings, perhaps only apparently smaller. I think it is just here the capital defect in the present method of treating chronic diseases of the lungs consists. When reduced to its ultimate conditions these disorders are assuredly due to perverted nutrition. To remedy this notable defect the stomach is properly plied with much nutritive material, dietetic and medicinal ; but this, to continue the figure, is but a lame attempt to close only one of the doors. There are important objects to be attained in the use of medicines.. But neither nutritive food nor skillfully prescribed medicines can do for a consumptive struggling with his disease all that ought and can be done ; these have no power to expand the chest, to restore the elasticity of its walls, inflate the collapsed and choked-up lungs, to relieve pulmonary congestion, and improve the capillary circulation by distributing the blood all through and over the body. It is very common to observe persons under the care of specialists in lung diseases of wide experience and great reputation, from the time that alarm is first excited, who—though under treatment from the very beginnings of the mischief—go from bad to worse ; but when special training is added, even after the disorder has made notable progress, the remedies which before seemed to be useless now exert the desired effects, and between the two substantial improvement is soon apparent, when either alone would be much less valuable.

PERMANENCE OF THE RESULTS.

It is well worthy of attention, as proving the curative influence of special training, when used with suitable medication, that it is successful in the cure of lung disease in climates that are believed to be favorable to its development, and unfavorable to its cure. If space permitted, I could cite the cases of a number of persons who contracted the disease while living in a district where raw, cold winds are very prevalent. They were there successfully treated, mainly by the measures here advocated, and have since continued in good health.

Now, if these persons had not recovered thoroughly, not only of the local lung trouble, but of the constitutional tendency thereto, they would certainly have relapsed, especially as they were constantly exposed to what is considered to be an unfavorable climate, before, during and after the treatment.

SPECIAL TRAINING AS A PALLIATIVE FOR INCURABLE CASES.

In this, as in every other serious disease, there is a point in its progress beyond which recovery is impossible. But if a sick man cannot get well, he is always glad to experience all the improvement of which his case will admit. If he cannot live ten, twenty or thirty years, he is thankful for, and glad to adopt anything that shall prolong his life five, three, two years, or even a single year. Now, although these sufferers cannot be cured, they can be quickly and materially benefited. Their blood is made to circulate more regularly, appetite improves, sleep becomes more refreshing, strength increases, the feeling of weariness with which they are oppressed is ameliorated. In fact, only a few days are required to mitigate their most distressing symptoms. Of course, the time comes when no farther improvement can be obtained.

I have, however, seen consumptives who were not expected to live many weeks, although under the care of eminent medical men, inprove as soon as special training was added, and live twice or thrice as many months.

SPECIAL TRAINING AS A PREVENTIVE.

For obvious reasons there is no disease in which an ounce of prevention is better worth a pound of cure than in consumption. Vaccination, sanitary science and improved methods of medical treatment have done much to prevent the occurrence altogether, and to limit the destructiveness of many diseases when they occur. Still, I am not aware that any effective move has been made to stay the great mortality caused by consumption. Yet I think much could be done in that direction, if people generally took the trouble to ascertain if their breathing capacity was up to the healthy standard; and particularly if those who know their lungs are weak, and whose family history shows that they may be congenitally predisposed thereto, would submit to timely training suited to strengthen the lungs and increase the vital volume, the mortality from consumption would be notably checked.

THE AUTHORS' PUBLISHING COMPANY'S
NEW BOOKS.

Evolution and Progress:

An Exposition and Defence. The Foundation of Evolution Philosophically Expounded, and its Arguments (divested of insignificant and distracting physical details) succinctly stated; together with a review of leading opponents, as Dawson and Winchell, and quasi-opponents, as Le Conte and Carpenter. By Rev. WILLIAM I. GILL, A. M., of Newark Conference, N. J. *The first volume of the International Prize Series.* THIRD EDITION. Cloth extra, imitation morocco, fine paper, 295 pp., 12mo., Price . $1 50

Each volume in this series was awarded a prize of *Two Hundred Dollars* in addition to copyright, in a competition which was open one year to the world, and where over three hundred manuscripts were submitted and read.

DESCRIPTIVE OPINIONS OF EVOLUTION AND PROGRESS

One of our most candid and thoughtful writers.—*Dr. Crane.*
He is a clear and strong reasoner.—*Cincinnati Christian Standard.*
A particularly strong argument.—*Evansville (Ind.) Daily Journal.*
It is ably written. Builds on philosophical principles.—*Brooklyn Union.*
The attitude of Mr. Gill, and his courage in maintaining it, are worthy of note.—*New York World.*
I rejoice in all attempts of this kind, made in a spirit like that which prompts your work.—*Herbert Spencer.*
His writings are marked by strong common sense, sound logic, and clear demonstration.—*Methodist Home Journal.*
It is a book of original thinking on one of the greatest themes.......A keen, thoughtful, vigorous volume.—*Golden Age.*
He strikes with no velvet glove, but with a steel-clad hand, dealing his blows with equal profusion and impartiality.—*New York Tribune.*
His effort is earnest, able and bold...... It presents, in all their naked strength, thoughts and arguments which will have to be met and answered.—*The Methodist, New York.*

Analytical Processes;

Or, The Primary Principle of Philosophy. By REV. WILLIAM I. GILL, A. M., author of "Evolution and Progress." *The Third Volume of the International Prize Series.* Cloth extra, fine paper, uniform with "Evolution and Progress," 450 pp., 12mo. Price $2 00.

A work which the committee cannot describe without seeming to exaggerate. It is marked by extraordinary depth and originality, and yet it is so clear and convincing as to make its novel conclusions appear like familiar common sense.—*From Report of Committee of Prize Award.*

It contains a vast amount of able and conscientious thought and acute criticism.—*Dr. McCosh, Prest. Princeton College.*

A specimen of robust thinking. I am very much gratified with its thoroughness, acuteness and logical coherence.—*Dr. Anderson, Pres't Rochester University.*

Ecclesiology:

A Fresh Inquiry as to the Fundamental Idea and Constitution of the New Testament Church; with a Supplement on Ordination. By REV. E. J. FISH, D. D. Cloth extra, fine paper, 400 pp., 12mo. . Price $2 00.

DOCTOR FISH disposes this volume into four parts.—I. The Fundamental Idea of the Church; II. The New Testament Church Constitution; III. Application of Principles; IV. A Supplement on Ordinaton—and addresses himself to his themes with the full earnestness of ability, clearness of logic, and conscientiousness of spirit which comprehensive treatment requires. As a "building fitly framed together," it is a fair-minded and standard contribution to the best religious literature of the Christian age.

The Beauty of the King:

By REV. A. H. HOLLOWAY, A. M., author of "Good Words for S. S. Teachers," "Teachers' Meetings," etc. Cloth extra, 174 pp. 12mo, $1.00; full gilt, beveled edges, $1.25.

A remarkably clear, comprehensive and intelligible exposition of the natural and spiritual causes, processes and effects of the birth, life and death of Jesus—a subject much discussed, yet not generally understood now-a-days.

Life for a Look:

By REV. A. H. HOLLOWAY, A. M. Paper covers, 32mo.

Price, 15 cts.

Earnest, cogent words, marrowy with the spirit of honest, old-fashioned Religion.

Christian Conception and Experience.

By Rev. WM. I. GILL, A. M., author of "Evolution and Progress," "Analytical Processes," etc.   Imitation Morocco 12mo.   .   .   .   .   Price, $1 00.

A fresh exposition and argument, practically enforced by a remarkable narrative of the conversion of a skeptic through this same argument. While it exhibits in parts the philosophic cast of the author's mind, its vivacious and lucid treatment will create for it a universal interest. This third work—in order of publication— by this fearless investigator, has, in large part, been written since his Trial before the Newark Methodist Episcopal Conference, under the charge of "Heresy," for writing his EVOLUTION AND PROGRESS, and it supplies abundant, fresh and vigorous thought-pabulum for the entertainment of heretics, critics, and Christians alike.

Resurrection of the Body. Does the Bible Teach it ?

By E. NISBET, D. D.   With an Introduction by G. W. SAMSON, D. D., late President of Columbian University, D. C. Fine English cloth, 12mo.   Price $1.00.

This is the careful work of an independent thinker and bold investigator. He strips away the trammels of hereditary prejudice, breaks the "old bottles" of unreasoning bias, and, with invincible logic, enters a field of research which had almost made a coward of thought. He begs no questions, makes no special pleadings, but meets the issue in its full front with such clean honesty and consummate ability that the book will interest and instruct every fair-minded reader, and charm and gratify every earnest student.

Reverend Green Willingwood ;

Or, LIFE AMONG THE CLERGY. By Rev. ROBERT FISHER. Silk cloth, ink and gold, beveled edges, full gilt. 12mo, $1.25.

With a resolute spirit and a knightly lance the Rev. Green Willingwood fights the battles of his brother clergymen.   His battle ground is in the midst of every congregation.   His armament is comprised of faithful work, hearty humor and delicate satire.   In short, Rev. Green Willingwood says and does precisely that which is wont to be said and done, but which, for obvious reasons, cannot be spoken from the pulpit nor accomplished directly in the pastorate.

Deacon Cranky.

By GEORGE GUIREY. Cloth extra, clear type. Price..$1.50

A bright and vigorous story in which every reader will readily recognize the familiar form of Deacon Cranky, whose strong points are superbly developed by Church Fairs, Choir troubles, Charity Contributions, Dorcas Society missions, religious Sleigh-rides and moral Necktie Parties, while the thread of the story retains vital earnestness, sharp characterization, and absorbing interest throughout.

Is Our Republic a Failure?

A discussion of the Rights and the Wrongs of the North and the South. By E. H. WATSON, author of "United States and their Origin," etc. English cloth, ink and gold, 12mo, 436 pp. . . . . . . Price, $1.50

In a spirit of genuine candor and unswerving impartiality.—*N. Y. Sun.*

It is fair, candid, impartial, the whole subject well treated.—HON. J. H BLAKE, *of Boston.*

I like the spirit of the book, its comprehensive patriotism, its liberal spirit, and its healing counsels.—HON. GEO. S. HILLARD, *author of "Franklin Readers," "Six Months in Italy,"* etc.

I read the manuscript with much interest—an interest belonging to the arguments themselves, but now increased by the perfection given to the form and style.—HON. MARTIN BRIMMER, *Boston.*

Lucid and just. The method of the argument, the facts on which it proceeds, and the conciliatory spirit which invests them, contribute to the book a value which cannot be too highly estimated.—GEN. JOHN COCHRANE.

The principles of American statesmanship which it asserts, must essentially prevail, unless we are so soon to fall from our original high plane of constitutional republicanism. I shall spare no exertion to promote the knowledge of such an able and impartial and statesmanlike compendium of our present political philosophy.—HON. JOHN QUINCY ADAMS, *Mass.*

Clearly expressed, and the argument is closely and ably maintained. The tone and the temper of the writer are beyond praise. They are as valuable as they are rare. They are those of a patriotic and philosophical observer of men The like spirit everywhere diffused among our people would make fraternal union as certain as desirable ; and if brought to the discussion of public affairs, would secure the adoption of wise and beneficent counsels.—HON. GEO. H. PENDLETON, *Ohio.*

Universe of Language.

I.—ITS NATURE. II.—STRUCTURE. III.—SPELLING REFORM Comprising Uniform Notation and Classification of Vowels adapted to all Languages. By the late GEORGE WATSON, Esq., of Boston. Edited, with Preliminary Essays, and a Treatise on Phonology, Phonotypy and Spelling Reform, by his daughter, E. H. WATSON, author of "Is Our Republic a Failure?" etc. Cloth extra, tinted paper, 12mo, 344 pp. Price $1.50

One of the great scientific labors of Mr. Watson's life was to segregate and systematize the universal elements of Language. His investigations were broad and comprehensive. Miss Watson has rounded her father's work with worthy zeal and eminent ability ; and the result, in this volume, is a unique and learned contribution to the permanent advantage and advancement of philology.

PRACTICAL THOUGHT.

Mercantile Prices and Profits;

Or the Valuation of Commodities for a Fair Trade. By M. R. PILON. Handsomely printed, 8vo., paper, 100 pp., ... **In Press.**

The author has brought broad experience and comprehensive research to bear upon his subjects. His style is terse and perspicuous. He uses the easy and concise language of an educated business man ; and, with wonderful art, invests every chapter with the grace and charm of a well-told story.

Monetary Feasts and Famines;

Labor, Values, Prices, Foreign and Fair Trade, Scarcity of Money and the Causes of Inflation. By M. R. PILON, author of "The Grangers." Uniform with "The Grangers,"—(*In Press.*) . .

What is Demonetization?

Ways to arrive at the Demonetization of Gold and Silver, and the establishment of Private Banks under control of the National Government. By. M. R. PILON, author of "The Grangers." FIFTH EDITION. 8vo., 186 pp., paper cover, . . Price 75 cents.

The work is interesting, and especially valuable to financiers.—*Jersey City Daily Journal.*

He gives expression to a good deal of sound financial principle.—*Louisville Daily Commercial.*

It is full of common sense......Valuable for its facts, its thoughts and its suggestions.—*Troy Daily Whig.*

Is written in an interesting and popular style and contains much useful information.—*Oakland, Cal., Daily News.*

The subject of the high valuation of gold and silver currency is fully discussed, and offers some new ideas worthy the attention of those interested in monetary affairs.—*Toledo Commercial.*

The author is a merchant who has extensively studied the currency problem. His hits are often sharp and incisive........Mr. Pilon would provide ample banking facilities for every city, town and village, with both stock and land security.—*Cincinnati Daily Star.*

........Discussing the currency question in an original, forcible and entertaining style. The author has brought together a great amount of varied information upon the whole subject of money......Those interested will find unquestioned ability in the author's handling of it.—*Baltimore Methodist Protestant.*

The Manuscript Manual: .

How to Prepare Manuscripts for the Press—practical and to the point. Paper, 26 pp., 8vo. Price 10 cents.

A most useful little companion to the young writer and editor.—*The South, New York.*

Gives excellent hints to intending writers.—*Cleveland Evan. Messenger*

The Race for Wealth,

Considered in a Series of Letters written to each other by a Brother and Sister. Edited by JAMES CORLEY. 12mo, 180 pp., paper . . . . . . Price 50 cents.

Shows how labor strikes may be prevented; how women may advance their political influence; how marriage may recover due regard in public opinion; the impossibility of enforcing total abstinence from strong liquors; and treats these and other topics of social and political economy in a clear style, making the work peculiarly attractive and impressive.

Aptly considered.—*St. Louis Christian.*
Of special importance.—*Cincinnati Gazette.*
Attractive . . . needed.—*Quincy Whig.*
Sensible, robust, sound.—*Hartford Courant.*
Clear, earnest, thoughtful.—*Phila. Nat. Baptist.*
Pleasant, intelligent, wholesome, useful.—*Zion's Herald, Boston.*
Simplicity in the arguments and the way of presenting them that is refreshing.—*Louisville Courier Journal.*

Author's Manuscript Paper.

Made from superior stock, in two grades, and sold only in ream packages. Each package warranted to contain full count of 480 sheets.

MANUFACTURED BY THE AUTHORS' PUBLISHING COMPANY.

AUTHOR'S MANUSCRIPT PAPER, 5¾ + 11, per ream . . . $1.00
AUTHOR'S MANUSCRIPT PAPER, 5¾ + 11, heavier, per ream , 1.25

NOTE.—When paper is sent by mail 50 cents per ream, in addition to price, must accompany order, to prepay postage.

It is only by making a specialty of this paper, manufacturing directly at the mills in large quantities, and selling exclusively for cash, that the demand can be supplied at these low prices. It is really nearly ONE HUNDRED PER CENT. cheaper than any other paper in the market.

It is ruled on one side, the other plain; is approved by writers and preferred by printers; and it has now become the popular standard paper for authors, contributors, editors, and writers generally.

☞ The A. P. Co. sell no other stationery.

A very convenient size, and at a low price.—*Publishers' Weekly, N. Y.*

The distinguishing feature of the Manuscript Paper is its convenient shape. The texture is neither too thick nor too thin, making it in every way a desirable paper for writers and contributors.—*Acta Columbiana, New York.*

It is especially useful for writers for the press, combining as it does good quality. with cheapness. The convenience of form is apparent to all who have writing to do. while it soon saves its price in postage.—*Essex County Press, Newark, N. J.*

Thousands of letters from well known authors, editors, and writers are on file in our office expressing the highest satisfaction with this paper, and thanking us for introducing it into market.

ÆSTHETIC THOUGHT.

Irene; or, Beach-Broken Billows:

A Story. By MRS. B. F. BAER, author of "Lena's Marriage," "The Match-Girl of New York," "Little Bare-Foot," etc., etc. *The second volume of the International Prize Series.* SECOND EDITION. Cloth extra, fine thick paper. 12mo.　.　.　.　Price $1 00.

Natural, honest and delicate.—*New York Herald.*
Charming and thoughtful.—*Poughkeepsie Eagle.*
Depicted in strong terms.—*Baptist Union, New York.*
Eminent.y pleasing and profitable.—*Christian Era, Boston.*
A fascinating volume.—*Georgia Musical Eclectic Magazine.*
Characters and plot fresh and original.—*Bridgeport News.*
With freshness, clearness, and vigor.—*Neb. Watchman.*
Delightful book.—*Saturday Review, Louisville, Ky.*

Lays open a whole network of the tender and emotional.—*Williamsport (Pa.) Daily Register.*

The unity is well preserved, the characters maintaining that probability so essential in the higher forms of fiction.—*Baltimore Methodist Protestant.*

There is a peculiar charm in the reading of this book, which every one who peruses it must feel. It is very like to that which is inspired in reading any of Hawthorne's romances.—*Hartford Religious Herald.*

Wild Flowers:

Poems. By CHARLES W. HUBNER, author of "Souvenirs of Luther." Elegantly printed on fine tinted paper, with portrait of the Author, imitation morocco and beveled edges, 196 pp., 12mo. *Just ready,*

Price $1.00. The same, gilt top, beveled edges, **$1.25**

As a poet Mr. HUBNER is conservative—always tender and delicate, never turbid or erratic. He evinces a strong love of nature and high spirituality, and brings us, from the humblest places and in the humblest guises, beauties of the heart, the life, the universe, and, while placing them before our vision, has glorified them and shown that within them of whose existence we had never dreamed.

Her Waiting Heart:

A Novel. By LOU CAPSADELL, author of "Hallow E'en." Cloth extra, 192 pp., 12mo. *Just ready.* $1 00.
A story of New York—drawn from the familiar phases of life, which, under the calmest surfaces, cover the greatest depths. Charming skill is shown in the naturalness of characterization, development of plot and narrative, strength of action and delicacy of thought.

Women's Secrets; or, How to be Beautiful:

Translated and Edited from the Persian and French, with additions from the best English authorities. By LOU. CAPSADELL, author of "Her Waiting Heart," "Hallow E'en," etc. Pp. 100, 12mo.

Saratoga Edition, in Scotch granite paper covers, 25 cents.
Boudoir Edition, French grey and blue cloths, . 75 cents.

The systems, directions and recipes for promoting Personal Beauty, as practiced for thousands of years by the renowned beauties of the Orient, and for securing the grace and charm for which the French Toilette and Boudoir are distinguished, together with suggestions from the best authorities, comprising History and Uses of Beauty; The Best Standards; Beautiful Children; Beauty Food, Sleep, Exercise, Health, Emotions· How to be Fat; How to be Lean; How to be Beautiful and to remain so, etc., etc.

Sumners' Poems:

By SAMUEL B. SUMNER and CHARLES A. SUMNER. With Illustrations by E. STEWART SUMNER. On fine tinted paper, 518 pp., cloth extra. Regular 12mo edition, $2.50 Large paper, 8vo, illustrated, full gilt, beveled edges...$4.00

Sparkling, tender and ardent.—*Philadelphia Book Buyer.*
Vivacity and good humor.—DR. OLIVER WENDELL HOLMES.
Brilliant and humorous, patriotic and historic.—*American Monthly, Phila.*
Equal to anything that is at all akin to them in "The Excursion."—*N. Y. World.*

The Buccaneers:

A stirring Historical Novel. By RANDOLPH JONES, Esq. Large 12mo, cloth extra, ink and gold. Paper $1. Cloth $1.75.

Is drawn from the most daring deeds of the Buccaneers and the sharpest events in the early settlement of Maryland and Virginia. It is so full of thrilling action, so piquant in sentiment, and so thoroughly alive with the animation of the bold and ambitious spirits whose acts it records with extraordinary power, that the publishers confidently bespeak "THE BUCCANEERS" as the most strongly marked and the best of all American novels issued during the year.

Cothurnus and Lyre.

By EDWARD J. HARDING. Fine English cloth, ink and gold, 12mo, 126 pp.... .... $1 00

Real poetic feeling and power.—*Am. Bookseller.*
Nobility not without sweetness.—*N. Y. World.*
Vigor which is quite uncommon.—*London Spectator.*
A unique and striking work.—*Boston Home Journal.*
Models of neatness and consideration.—*N. Y. Commercial.*
Has created a sensation in Eastern literary circles.—*Chicago Herald.*

Spiritual Communications.

PRESENTING A REVELATION OF THE FUTURE LIFE, AND ILLUSTRATING AND CONFIRMING THE FUNDAMENTAL DOCTRINES OF THE CHRISTIAN FAITH. Edited by HENRY KIDDLE, A. M. Cloth extra, 12mo, 350pp.........$1 50

This is no ordinary book; indeed, it contains the most startling revelation of modern times. What the eminent educationist and author commenced as an investigation into certain remarkable psychological phenomena, brought to his notice in a very singular manner, has culminated in the wonderful record presented to the public in this volume.

STARTLING REVELATIONS! FACTS ATTESTED!

While it is a most important addition to the literature of Spiritualism, the growth of which is, perhaps, the most amazing fact of our times, it is far more than this. It comes as a tocsin of the New Jerusalem, an evangel of " Peace and good will toward men,' a herald of the " Resurrection of the world," and the " Second Coming of Christ." The internal senses are opened! The departed return! Dead and living clasp hands! Men and angels speak together! The celestial curtain is rolled back! The natural and spiritual worlds stand face to face!

This is no exaggeration. The book attests it all as a *reality;* for it is no mere speculation, but the record of *living facts;* the logical evidence to support which may be briefly stated as follows:

1. That it comes through the wonderful gifts of one of the purest, simplest, and most truthful of minds

2 That it has received the careful investigation of a man of ripe intellectual culture, distinguished for scholarly attainments, sound practical common sense, and purity of personal character, whose whole life has been a rigid mental training, and whose successful career in the field of education has reflected the highest credit upon himself, and has brought honor to the City of New York both at home and abroad

3. That the teachings and tendency of the book are spiritually or religiously of the purest and sublimest character. No man, whatever may be the characteristics of his mind or religious faith, can ever rise from the candid perusal of this book without becoming a purer and better man.

4. That the internal evidence comprised in the communications themselves,—so vast in their scope, so various in their style, so startling in their statements—is so indubitably plain that he who runs may read and understand. To no book ever written, except the Sacred Scriptures themselves, are the well known lines of Scott so applicable :

" Within that awful volume lies
The mystery of mysteries.;
* * * * * * * *
And better had they ne'er been born
Who read to doubt, or read to scorn."

ILLUSTRIOUS MESSENGERS FROM OTHER WORLDS.

Here, with the voice of *inspiration*, speak the spirits of the departed—the *illustrious of earth*, –Shakespeare, Byron, Shelly, Bryant, Poe, Washington, Lincoln, Bacon, Newton, etc., etc ; the *personages of sacred history*—Moses, the Prophet of old, St. Peter, St Paul, St. Augustine, Pontius Pilate, etc.; Christian *ministers of various denominations*, Luther, Calvin, Bishop Ives, Archbishop Hughes, Pio Nono, Dr Channing, Theodore Parker, Bishop Janes, Dr Muhlenberg, etc., and the *Seer* Swedenborg Here speak to us the spirits of blissful spheres; and here also the spirits of the sinful and erring come, and tell their sad experience, as a lesson to mankind

The bright spheres and the dark world are here, in part unveiled to mankind, so that they may choose between them. But in every page of this wonderful volume, the infinite goodness and mercy of God and the love of our Saviour Christ are shown with flashes of heavenly light.

No notice can give any adequate idea of the character of this book, which, it is not saying too much to declare, contains an *evangel* that is destined to travel the world over.

THE ENCHANTED LIBRARY.

COMPRISING

Stories, Sketches, Travels, Adventures, etc., etc.,

FOR YOUNG FOLKS.

Handsomely printed from large type, with elegant and substantial Cloth binding, in uniform style, Child's quarto.

Only such works as are thoroughly excellent in all respects, and by authors who are experienced and eminently skillful in writing books for **Young Folks** especially, will be offered in this Series.

The Publishers will conscientiously endeavor to make every volume of the **Enchanted Library** a pure delight and an enchanting visitor to every growing Family.

Queer Little Wooden Captain AND THE LITTLE LOST GIRL. By SYDNEY DAYRE. 152 pp..................90c.

"The Little Wooden Captain" includes Grandma's story of the "Menagerie on the Farm," where they tried to make the Calf an Elephant, the Cat a Monkey, the Rooster an Ostrich, etc.; the Christmas Frolic of the Broom, Tongs, Shovel, Poker, Kettle, and Teapot; the Little Wooden Captain's funny story of himself and how he got entangled with a dozen other little Wooden People with which an old Clock-Maker had ornamented his products.

"The Little Lost Girl" was carried off in the arms of a frightened Nurse during the war; and her experience makes one of the most beautiful, pathetic and delicious little stories. From beginning to end this story is exquisitely interesting to both Girls and Boys.

Harry Ascott Abroad. By MATTHEW WHITE, JR., 115 pp. ....60c.

This is an American boy's experience abroad, charmingly narrated in genuine boyish spirit and with a naturalness that fascinates while it instructs. He hunted out the wonders which boys delight in, and describes them with continuous interest.

The peculiar sights of Hamburg, Frankfort-on-the-Maine, Heidelburg, Baden-Baden, Stuttgart, Nuremburg, Dresden, Berlin, Cologne, Bâle, Berne, Interlaken, Lucerne, Paris, London, etc., contribute the "brother of a baron," fairylike castles, eccentric acquaintances, royal turnouts, curious student caps, chairs that played tunes when sat upon, glittering hussars in white boots, horse-cars made wrong side out, and thousands of other oddities and pleasing curiosities for young eyes.

OTHER VOLUMES FOLLOW SHORTLY.

Common Sense; or, First Steps in Political Economy.

A Manual for Families, Normal Classes and Schools, with an Appendix for Teachers and Students. By M. R. LEVER-SON, Dr. Ph., A. M., cloth, 12mo, 215 pp...........$1.25

Excellent.—*N. Y. Nation.*

Well suited to its purpose.—*Indianapolis Journal.*

Simple and clear, useful and interesting.—*N. Y. Mail.*

The best authorities are followed under all the heads.—*San Francisco Bulletin.*

The author is certainly a man of extensive study.—*Denver Rocky Mountain News.*

I commend it.—Prof. J. H. HOUSE, *State Normal and Training School, Cortland, N. Y*

THE SATCHEL SERIES.

COMPRISING

Story, Romance, Travel, Adventure, Humor, Pleasure.

BY POPULAR AMERICAN AUTHORS.

Printed from bold, clear type, on good paper, with bright, cheerful
pages, and neatly bound in paper covers.

WELL RECEIVED! SELL WELL EVERYWHERE! A GREAT HIT!

THE BOOKS OF THE "SATCHEL SERIES" ARE:—

Handy little volumes.—*Philadelphia Record.*
Bright things by American authors.—*Whig, Quincy, Ill.*
Instructive as well as agreeable.—*New Bedford Standard.*
They are not of the trashy dime novel class.—*Glens Falls Messenger.*
Really of a lively and spicy character.—*American Monthly Magazine.*
Gotten up in a fresh style and printed in plain type.—*Pittsburg Leader.*
Cheap, convenient, and by popular authors.—*Epis. Methodist, Baltimore.*
Bright and breezy, and above all, pure in sentiment.—*Boston Transcript.*
They deserve well of the reading public.—*Illustrated Christian Weekly.*
Acknowledged by all to be bright, elegant and charming. There is
nothing trashy about them.—*Journal, Somerville, Mass.*
Bright and brief—just the books to read in the cars, at the seashore, or
during leisure hours at home.—*Chronicle. Farmington, Me.*
The brightest and best brief works by American authors who are well
known to the reading public. They have proved very popular, particularly
as summer travelling companions.—*Boston Home Journal.*
Complete in themselves, and interesting. They fill a void, for heretofore
cheap literature has been of the flashy and sensational kind, injurious not
only to the readers, but to the whole community, because of its effect. A
pleasing feature of the volumes issued by the AUTHORS' PUBLISHING COM-
PANY is that, while they are not dry or insipid, they have a moral tone and
effect.—*Fall River Herald.*

NEW VOLUMES ISSUED WEEKLY.

Lily's Lover;

OR, A TRIP OUT OF SEASON. By the
author of "Climbing the Mountains," etc. 135 pp...... 35 cents.

Is entertaining.—*Cin. Times.*
Written in a pleasing style.—*N. Y. Mail.*
A genuine love story.—*Epis. Methodist.*
Lily is the flower of the family.—*Epit.
of Literature.*
Really a well told story of love's trans-
formation —*Advance.*
Those who would pass an hour or two
pleasantly, should read it.—*Christian, St.
Louis.*
Is a very sweet and pretty story of sum-
mer-time romantic adventures among the
green hills and silvery lakes of Connecti-
cut.—*Boston Home Journal.*

Rosamond Howard.

A story of fact and fancy. By
KATE R. LOVELACE. 112pp. 25 cts.
Cloth extra............. 60 cts.

Abounds in quiet pathos.—*Philadelphia
Record.*
Has excellent points, and the heroine
has a well sustained character throughout.
—*Epitome of Literature.*
The tale is told with an evident sym-
pathy with all that is good in woman, and
makes the daintie kind of summer reading.—*N. Y. Mail.*
A modest, touching and attractive story
—pathetic and beautiful The story is
well unfolded and excellently managed.
It begins interesting, continues and ends
interesting.—*The Christian, St. Louis.*

Nobody's Business.

By the author of "Dead Men's Shoes," "Heavy Yokes," "Against the World," etc. 128pp....30 cts.

Spicy.—*Boston Contributor.*

A charming summer book.—*Epitome of Literature.*

An admirable book to take to the seashore.—*Phila. Record.*

Full of lively sallies and bright hits.—*Baptist Weekly, N. Y.*

Amusing incidents pleasantly related.—*Mo. Review Cur. Literature.*

A more charming book for sheer amusement we have seldom met with. *Mail.*

A great provocative to wakefulness and pleasure in traveling.—*Providence Press.*

We have not enjoyed anything so much for many a day.—*Potter's American Monthly.*

A piquant and spicey story, exceedingly entertaining, well written and full of snap.—*Meth. Pro't, Balto.*

It is sprightly, capitally told, sympathetically interesting, and highly amusing.—*Boston Home Journal.*

A breezy, spicy, rollicking story, admirably told and of unexceptional moral tone.—*Ill. Chris. Weekly, N. Y.*

Affords more hearty laughs than can be gotten out of the best comedy. The style is polished and spirited, the wit piquant, and its common sense sound as a dollar.—*Vicksburg Herald.*

Is a bright, sparkling extravaganza, and told in a very sprightly manner a good book to drive away the blues. Creates the impression produced in a sick-room or a sedate and solemn company by a bright, breezy, cheery, pretty little woman.—*Christian Union.*

Our Peggotties.

To all women who appreciate the situation. By KESIAH SHELTON, author of "Heights and Depths of Ambition," "Humbugged," etc. 106 pp................25 cts.

Is well adapted for an idle hour.—*Mail.*

Remarkably witty little book.—*Clyde, N. Y., Times.*

Sketches of New England life cleverly written.—*Balto. Gazette.*

All housekeepers will appreciate it.—*West Meriden, Ct., Recorder.*

It was a very fruitful experience and is rich reading.—*Providence Press.*

An amusing account of trials and tribulations with "help."—*Phila. Record.*

A tale of woman's woe in the matter of servants—semi-humorous.—*N. Y. Post.*

Semi-humorous account of a New England woman's experience.—*Hartford Post.*

Affords a dollar's worth of recreation for a winter's evening.—*Woonsocket Reporter.*

Deeply interesting; and above all pure in sentiment.—*Farmington, Me., Chron.*

A pleasant companion while traveling or in the drawing-room at home.—*Odd Fellows' Register, Providence.*

Our Winter Eden.

PEN PICTURES OF THE TROPICS, with an Appendix of the Seward-Samana Mystery. By MRS. GENL. WM. LESLIE CASNEAU, author of "Hill Homes of Jamaica," "Prince Kashna," etc., who was lost on the *Emily B. Souder*, December, 1878, en route to her beautiful "Winter Eden" which she so charmingly describes in this work. 112 pp........30 cts.

Bright, readable, and has value beside.—*N. Y. Mail.*

A bright and vivid description. It is written with a jaunty pen.—*Balt. Gazette.*

A pleasantly written book of travel and life in the West Indies.—*Buffalo Courier.*

Depicts life in that climate in all its beauty and attractiveness.—*New Bedford Standard.*

An interesting book, giving pen pictures of the Tropics in a charming manner.—*Boston Home Journal.*

A pleasant description of the peninsula of Samana which our government came so near purchasing.—*Boston Transcript.*

Describes the pleasure of life in the West Indies in glowing terms.—*Cin. Gaz.*

Relates to the Island of Samana, and as the author has lived there for some time, she is perfectly familiar with it, and gives many details of interest. It is well written.—*Boston Globe.*

By the wife of the U. S. Plenipotentiary to the Dominican Republic during the Pierce administration. Makes public for the first time the real cause of the failure of Secretary Seward's Samana scheme.—*Whig, Quincy, Ill.*

A Story of the Strike.

SCENES IN CITY LIFE. By ELIZABETH MURRAY. 125 pp..30 cts.

Vivacious tale.—*N. Y. Mail.*

Sound sense—will subserve a good purpose.—*San Fran. Chron.*

A pleasant story, inculcating economy, thrift, home-virtue, and honest industry as the basis of well-being and happiness. *Sunday-School Times, Phila.*

This is one of the few books called out by the great railroad strike. The style is crisp and taking, and the book healthy and captivating.—*Temperance Union.*

Is a faithful delineation of city life among the high and low. Without being sensational or trashy it is lively and interesting.—*Providence Town and Country.*

A most timely book—a series of pen-pictures in which the philosophy of strikes and poverty and riches is illustrated. It tells the story completely, and with an aptness no political economist can excel, while it is perfectly level to the comprehension of any reader. Should be in every family and read by every man, woman and child of to-day.—*Meth. Protestant.*

Prisons Without Walls.

A Novel. By KELSIC ETHERIDGE.
Paper. 97pp............25 cts.

The heroine, Egypt, is a glorious being.
—*N. Y. South.*

Has the curiosity-exciting tendency.—
Boston Beacon.

The characters are finely wrought up.—
Williamsport (Pa.) Register.

The interest grows and retains attention
to the end.—*N. O. Picayune.*

Is written in easy style and is sensational enough to sustain its interest to the
end.—*Providence (R. I.) Town and Country.*

Short, sententious, marrowy, and spiced
with episodes. Has a warm Southern
aroma of orange and magnolia blossoms.
—*Baltimore Meth. Protestant.*

Of rare beauty and power in its vivid,
life-like picturing of men and places.
Through such artistic touches of skill and
strength we are wafted in thought as we
follow the hero and heroine through the
mazes of the old, old story.—*Ladies'
Pearl, St. Louis.*

Strange, weird story. The style is peculiar, and has a wonderful fascination
about it. We feel, while reading it, that
the strongest bars which hold us prisoners
are those which fate casts about us, whose
iron grasp we cannot unloose; that the
walls built highest about us are those
which neither our will, nor our despair,
nor our unutterable agony, can batter
down.—*Kansas City Times.*

Traveller's Grab-Bag.

A Hand-Book for utilizing Frag-
ments of Leisure in Railroad
Trains, Steamboats, Way Stations
and Easy Chairs. Edited by an
OLD TRAVELLER. 110 pp..25 cts.

Full of spice and fun.—*Baltimore Meth.
Protestant.*

No traveler should be without it.—*N. Y.
Forest and Stream.*

Teeming with rollicking humor, and a
kind of satire that will be enjoyable.—
Pittsburg Commercial.

There are many good stories in this
book; some exciting and interesting,
while none are by any means dull.—*Star
Spangled Banner.*

It has three separate elements as unlike
as vinegar, aloes and honey. It is an odd
book, the design original, and is a grab-
bag literally in which to plunge the mind
by a glance of the eye.—*Kansas City
Times.*

Is a handy little volume of short, interesting original stories. Many a weary
moment may be beguiled away by the fatigued traveler, provided he takes the
"Grab-Bag" with him.—*Epitome of Literature.*

Bonny Eagle.

A Vacation Sketch—the humor of
roughing it. 121pp.......25 cts.

Exceedingly entertaining. — *Brooklyn
Times.*

Rich, racy and entertaining.—*Quincy
(Ill.) Whig.*

Some of the experiences are exceedingly ludicrous.—*Epitome of Literature.*

The style is fresh and graphic, and the
humor and satire are keen and pure.—
Boston Home Journal.

Funny; conceived and executed in great
good humor. Bright and entertaining.—
Chicago Sat. Eve Herald.

Will be read with interest and amusement, and many a tear (of laughter) will
be shed over its all too few pages.—*San
Francisco Post.*

Curious and ludicrous experiences graphically told with a naive humor and delicate satire. It is a fresh and spicy book.
—*St. Louis Herald.*

Very spicy, humorous, satirical. Particularly interesting. Delightful hours
amid forest scenes of beauty and retire-
ment.—*Hebrew Leader, N. Y.*

The relation of the haps and mishaps,
mild experiences of "roughing it" under
canvas, their fraternal intercourse with
the Dryads and Hamadryads of the grove,
and the varied incidents, is given in grace-
ful language. A friendly expedition has
rarely been chronicled in better style.—
The South, New York.

Voice of a Shell.

By O. C. AURINGER, 180pp.40 cts.

A fine collection.—*Pittsburg Leader.*

Well written —*Schoharie Republican.*

Contains many fine lines.—*Bal't. Gaz.*

There is no lack of fire and passion.—
Literary World.

It is delicate, beautiful and grand.—
Sandy Hill Herald.

Most of them written while at sea.—
Glen's Falls Messenger.

There is much in the book that is really
fine.—*Glen's Falls Republican.*

Much poetic feeling, and an absorbing
love of the sea —*Herald, Chicago.*

Full of peculiar interest, grandeur and
tenderness.—*Boston Home Journal.*

Poetic merit. Most of them are short,
and present a pleasant variety.—*Troy
Whig.*

A book of poems, having—as their col-
lective title implies—a salt-water flavor.—
Syracuse Herald.

Exceedingly pleasing, in their sweet
delicacy of thought. Poetic and home
like.—*N. Y. Mail.*

A touch of real poetic feeling and originality. The author's feeling for the sea
is evidently an intense one.—*American
Bookseller.*

To all lovers of the sea, and to all who
linger by its sounding shores, nothing can
be more entrancing than the pages of this
beautiful little volume.—*Altoona Tribune.*

Who Did It?

By Mark Frazier. 137pp. 30 cts.

An excellent theme.—*N. Y. Post*

Deeply interesting.—*Farmington, Me., Chronicle.*

In these times when grave-robbing is very common it is quite to the point.—*Whig, Quincy, Ill.*

A story with a suspected crime for the basis of its mysterious plot—well told and absorbing—*Cin. Gazette.*

A sensational story in which a case of catalepsy so closely resembles death that the girl is declared dead.—*New Bedford, Mass., Standard.*

Is a thrilling story, and he who commences reading it will never stop until his eyes have glanced over its last page—*Cycle, N. Y., Times.*

Is a deeply interesting story of a beautiful girl. Around the terrible incident of being buried alive is woven a romantic story.—*Boston Home Journal.*

Portrays the dangers of premature burial. Serves a good purpose in directing attention to the necessity of absolute certainty as to death before interment is allowed to take place.—*Boston Traveller.*

It would be difficult to find a more excitable and thoroughly readable story. The history of the narrative claims an origin in south-eastern New Hampshire, but the scene is thrilling enough to have been laid in a far more eventful country. It is just the book for an hour's reading on a winter's night.—*Mirror and Amr'n., Manchester, N. H.*

Earnest Appeal to Moody.

A Satire. 34 pp. 10 cts.

A clever poem—the hits well taken and to the point, and will be appreciated by many as the names are outlined sufficiently as to be readily recognized.—*Epitome of Literature.*

The references to Kingsley, McLoughlin, Kinsella and the other crooked "K's," Bob Furey, Bill Fowler and the rest will prove quite amusing, especially as it represents them as amenable to reform and to be conscience-stricken by Mr. Moody's preaching.—*Brooklyn Times.*

Only a Tramp.

By the author of "Alone," "Eone," "Through the Dark," etc. 212pp. 50 cts.

An exceedingly picturesque story, with a strong and strangely fascinating character, in the person of the adopted daughter, or kidnapped protege, of a Tramp, for its heroine. The Eyes of royal queens rarely furnished such heroic and beautiful material for romance as the author has here cunningly and ably drawn from the life of this poor little girl-tramp.

Bera;

Or, the C. and M. C. Railroad. By Stuart De Leon. A novel 169pp. 40 cts.

Well learned in his books, and fresh from the schools of many languages, was young Greek Lyk when fate turned his thoughts to love, and chance directed his steps to the Railroad shops of a far-away village in the Northwest. And here —with fantastic blending of the quaint, strange characters who open the streets of frontier towns—the principal action of this well-told story transpires. Its vivid and swiftly-moving scenes are bright and refreshing, like sunlight down the road over which the record runs.

Poor Theophilus;

And the City of Fin. By a Well-Known Contributor to "Puck." 99pp. 25 cts. Cloth extra. 60 cts.

A love story, tenderly touching, with much fact and little fancy; together with an oddly amusing and quaint conceit pertaining to the wonders of the sea, which is only the more interesting for the little fact and much fancy which it contains.

How It Ended.

By Marie Flaacke. 103pp. 25 cts.

A story, with a glow of southern climes and the perfume of orange groves about it. A real gem of a little book, in the inspiring influence which pervades the glowing landscapes and shady nooks of its well-framed pen-pictures; strong in its delicacy, love, and tenderness.

Glenmere.

A Story of Love *versus* Wealth. 112pp. 25 cts.

With well defined action, excellent characterization and thoroughly sustained interest, this compact story is attractive in both plot and purpose; and it is withal, forceful with the healthful tone of the great Northwestern country and people, whence its scenes and animation are drawn.

Our Smoking Husbands, AND WHAT TO DO WITH THEM. By HARRIET P. FOWLER. 47pp............10c.

This is a bright and suggestive little sketch, where good hits, good hints and good common sense are most agreeably blended for the reader's entertainment.

The story of "Our Smoking Husbands" is naively woven about

I. Mrs. Bird, Bride No. 1.

II. Mrs. Everett, Bride No. 2.

III. Mrs. Hammond, Bride No. 3.

The following volumes of the "Satchel Series" were put in press too late to be indexed in the current edition of this Catalogue.

Ninety Nine Days. By CLARA R. BUSH, —pp..—c.

A love story, well drawn, and written with much spirit and animation. Will have a large sale.

Spiders and Rice Pudding. By FRANCES G. STEVENS, —pp....................25c.

A very pleasing, compact and pretty story — fascinating and striking.

The attention of the TRADE is asked particularly to the "SATCHEL SERIES" as popular and fast-selling books.

Newsdealers and Railroad agents find them the most active and the most profitable stock they can handle.

Everybody likes them.

Doing a large business with this Series, and printing in very large quantities, we are prepared to make EXTRA-SPECIAL DISCOUNTS on these books when ordered in quantities.

'Twixt Wave and Sky.

A Novel. By Miss FRANCES E. WADLEIGH. Large square 12mo, cloth extra, ink and gold, 261pp. $1.25

This is a story of strong plot and rapid action, with a thoroughly natural and clever set of characters. The situations are exciting and novel without being sensational. The narrative is put in a curiosity-exciting form; the spirit and purpose are noble and pure throughout, and the climax is highly romantic.

In all respects and parts it is a story of very marked ability, and fully entitles the accomplished author to an immediate and distinguished rank among the ablest American writers of fiction.